THE FUSION AGE

MODERN NUCLEAR FUSION REACTORS

RICHARD DINAN

Acknowledgments:

We are grateful to the following companies, institutions and individuals :

EUROfusion for photos.

Dr James Lambert – Operations Manager at Pulsar Fusion for his research in helping the creation of this book.

Professor Stephen Cowley – Former CEO of the UK Atomic Energy Authority for his time and showing me around the JET and MAST facilities at the Culham Centre for Fusion Energy

Professor Ian Chapman - CEO of the UK Atomic Energy Authority for hosting meetings with me at the JET and MAST facilities at the Culham Centre for Fusion Energy

Professor Dennis Whyte - Director of the Plasma Science and Fusion Centre at MIT – For his time with Dr James Lambert of Applied Fusion Systems

ITER Organisation for hosting members of the Pulsar Fusion Team at their 2015 conference.

To the UK Government for its financial support of Pulsar Fusion and assistance from '

Business is Great - Great Britain'

Orders: Please contact info@appliedfusionsystems.com

ISBN 978-1-9998045-1-0

First published 2017

First edition

The Fusion Age - Modern Nuclear Fusion Reactors

Cover design by Lucy Ormiston

Inner page layout design by Sophie Lett

Printed in Great Britain by Ashwyk Ltd

Published by Applied Fusion Systems Ltd
D5 Culham Science Centre, Abingdon, Oxfordshire, OX14 3DB

FOR COSIMA

PREFACE

There is only one equation you need to know regarding fusion as far as I am concerned:

Funding = Fusion

You cannot expect people to invest in what they do not understand but considering energy markets turn over $7 trillion annually and nuclear fusion will one day be the planet's dominant power source, fusion is worth learning about. The motivation behind this book is to provide a practical guide to modern Tokamak reactors for non-physicists. Much existing content is created by scientists for scientists and is rich with equations that will quickly disinterest people whose resourcefulness may lie in other fields. Fusion development is calling out for the financiers and entrepreneurs of this world too.

Sitting on my desk as I write is a 38Kg Iron Meteorite. It is about the size of a football although kicking it would be extremely painful. The reason for it being there is not a decorative one. I left school at an early age to go about starting my first business. I didn't really care what it was, I just wanted to start out in life and not spend any longer looking out of the classroom window whilst fortunes were being made and lost as I was confined to tasks that would ultimately benefit me not at all, as far as I could see.

My first business was a student magazine which turned out to be very exciting but rather expensive. It demanded a lot more work that I had initially anticipated. I learned early the lesson of never entering into a business endeavour that you do not love. For those who love it, will always be ahead of you. From then on, all my financial endeavours were in pursuit of what I have always loved which is building (and taking apart) technology.

In 2013, whilst looking for rare and precious materials for a wearable technology company I am involved with, I tasked myself to see if it were possible to machine Iron Meteorite into shapes that could be worn and incorporated into our products. The problem was, sourcing large enough meteorite samples was extremely difficult and the 'World experts' advised me it was a fruitless mission as large enough samples are so very rare and even if found, there are only a few people on earth who can machine them. Today our company has one of the largest collections of meteorite in Europe and we have achieved some World firsts with underwater spark erosion techniques and the intricate machining of them. This was the reason I spent many years devoted to the study of iron meteorites and more importantly for this book, where they came from.

13.75 billion years ago, an event took place that started our universe, which was at the very beginning no larger than an atom but extremely hot and dense. It immediately started to expand. Almost right away, the Higgs field kicked in, giving what at the time were just quarks, their mass and a period of expansion began. After 400 million years, the first stars appeared and the universe had cooled enough for light to travel through it. These stars started to 'cook' hydrogen and fuse it into helium. As this first generation of stars started to run out of fuel, they began expanding, increasing their temperatures and cooking helium into carbon; and then on into heavier elements until at the end of their lives, in their hottest, final stages, Iron.

The Widmanstätten pattern, this can be seen by applying nitric acid to iron meteorite.

The iron in a meteorite contains a blending of iron and nickel in a visible arrangement called the Widmanstätten pattern. It captures the state of the two metals as they were at the death of a star long before our planet was formed. The iron is left to cool down at zero kelvin in space, for billions of years preserving this pattern. Sometimes, this iron meteorite can fall on Earth - and such is the piece I have in front of me now, 4.6 billion years old and the crudest, yet most beautiful evidence of nuclear fusion I know.

The reality is: all elements heavier than Hydrogen and Helium were cooked in stars long ago. Everything around you is literal evidence of nuclear fusion and if that weren't enough, it is all lit by a mighty fusion reactor in the sky for the eye to see.

Nuclear fusion is everywhere; it is the universe's chosen element creator and energy source. We have always followed the stars for direction. It is about time we finally learned to emulate them and solve the world's energy needs for good.

CONTENTS

CONTENTS

CHAPTER 1

WHAT IS NUCLEAR FUSION?

For over 50 years, scientists have been chasing the elusive possibility of harnessing the power of the Sun on Earth. Nuclear fusion, the process by which the nuclei of two lighter atoms are fused together to make one heavier nucleus, generates phenomenal amounts of energy - indeed, it powers the stars. It is clean, safe, and could solve the problem of our energy shortage and reliance on fossil fuels, which are driving up climate change - a threat to our very existence.

The Sun is not a solid, liquid or a gas, but a fourth state of matter, known as plasma. At the centre of the Sun, where temperatures reach 15,000,000^{o}C, conditions are too hot for electrons to stay attached to nuclei. Instead, hydrogen electrons, protons and neutrons move around within the plasma at extraordinary speed. The intense heat and pressure causes the hydrogen nuclei to fuse together, producing helium nuclei, and generating phenomenal amounts of energy in the process. This perpetual release of energy allows more nuclei to fuse, thus making for a continuous, gigantic, thermonuclear reactor in the sky.

As we know, the Sun produces enough energy to bring light and warmth to our planet, and it has therefore been a long-held ambition of the scientific community to harness the potential of nuclear fusion energy on Earth. Indeed, it is a long-standing joke in the energy sector that the world is just 30 years away from achieving nuclear fusion power - it always has been, and it always will be.

Achieving nuclear fusion itself is not the problem: humans have been able to trigger thermonuclear reactions since the 1950s, during the era of the first hydrogen bombs. The challenge is in harnessing the energy released by nuclear fusion. In order to do so, scientists have been trying for decades to recreate the conditions of the Sun within a lab. It is here that the challenges lie: to generate nuclear fusion in a reactor, you need to mimic conditions found in deep space. This is hugely expensive and uses enormous amounts of energy in itself; one of the major (and, at the time of writing, insurmountable) hurdles to date has been in achieving a net gain in energy from the nuclear reaction (known as 'energy-gain fusion'). In 1997, a major milestone was achieved when the Joint European Torus (JET) reactor produced 16MW of power at the Culham Centre for Fusion Energy in Oxfordshire. However, it took 25MW to power the reaction.

In order to achieve fusion in a lab setting, three conditions must be met: (1) extremely high temperatures, in the region of 150,000,000°C; (2) the plasma must be sufficiently held and confined as it expands; (3) the plasma must have particle density which is sufficient for collisions to occur.

If achieved, nuclear fusion represents the holy grail of the energy world: it is clean, safe and carbon-free. It involves zero fossil fuels: its primary fuel, hydrogen isotopes, can be found in normal seawater, which is abundant. We're talking here about an energy source that would combat climate change, enhance the world economy and make energy plentiful rather than scarce. Nuclear fusion doesn't produce the same nuclear waste that has made traditional nuclear power so controversial (which, incidentally, is fuelled by the process of nuclear fission - or the splitting apart of atomic particles). It has implications for space exploration and for defence, particularly for the countries that manage to develop it first. It would enable the use of fossil fuels to become a thing of the past. There's a reason so many entrepreneurs and investors are backing it. Put simply, it's the future.

During World War 2, there was a huge drive to create the nuclear bomb. This was using enriched, extremely heavy and unstable, radioactive elements like uranium and plutonium to create a chain reaction that would devastate the World.

They should have called these bombs 'Uranium bombs' or 'Plutonium bombs' as these are the materials responsible for the destruction. Nuclear is just the term for any processes involving the change in the nucleus of an atom. You and I are made up of billions of atoms, each with a nucleus. Our sun is a nuclear reactor and that's a pretty good thing I think you will agree. However, since the nuclear bomb, the word nuclear brings to mind thoughts of terrible destruction, death and the end of the World.

Chemical reactions similarly can be good or bad. You have bleach and acids and then you have fairy liquid and medicines. We must stop associating the word nuclear with bombs and all out, total destruction. A glass of orange juice is nuclear as is everything you see and touch in your life including this book. It won't blow up if you drop it, honest.

The nuclear bomb received huge amounts of funding like almost no other technology at the time. 'Development without budget'. Thus, our ability and understanding of this side of the technology matured quickly. Because of the 'nuclear bomb' we learned how to handle very nasty, radioactive materials and we also discovered other uses for them.

The difference between today's nuclear power stations and nuclear fusion

Very heavy elements like enriched uranium emit heat in the form of 'neutrons'. We will get to know these a little better later in this book. Once scientists had mastered the art of nuclear destruction as a priority, they started to consider using this enriched material to generate power. Heat from radioactive decay can be used to boil water and create steam to drive turbines and generate electricity.

Nuclear power stations today still use this method of enriching extremely heavy elements and capturing energy in the form of heat as they decay into lighter elements. The term for this process is 'Nuclear Fission'. Fission worked so reliably that it is now the only power source able to take over from coal today. The UK's National Grid, at a meeting in London, informed me that in their opinion, only two power sources are capable of consistently balancing the grid in times of heavy energy demand; coal and nuclear fission.

Fission plays such a key role in powering our cities today because it is ultimately safe, extremely powerful, reliable and relatively clean despite it involving the storage of long-lived radioactive material and the associated costly disposal of it.

This is the reason that the United Kingdom has just commissioned the £18 billion fission facility at Hinkley point C in Somerset, South West England. Hinkley C will create a 20-35% fixed return for its largely Chinese investors at the expense of the UK tax payer. The government has agreed through a CFD to purchase energy off Hinkley at £112 per megawatt hour. (MWh). This is not a particularly popular deal but as a nation, we absolutely need this energy.

At it's heart, fission involves handling very nasty materials and as a result it has had its problems. The fall-out at Chernobyl was the result of poor safety standards that would never have been permitted in the UK.

Hinkley Point. The new plant will be built next to the existing facilities.

An event like this would be extremely unlikely now. Not only does fission require the creation and storage of enriched radioactive materials, which must be protected by men with dogs and guns from falling into the wrong hands; but once the power has been obtained from the decay heat, we then must bury the remaining waste product for thousands of years in very deep holes to protect people from it. Fission receives a fair amount of public opposition but as an alternative to coal, it ultimately works out to be overwhelmingly preferable for the overall environment.

To power a city for a year would take burning a football stadium filled with coal in comparison to one single pick-up truck of fission rods that would be needed to generate the same amount of energy.

There is, however, a far better way still of obtaining energy from nuclear. One that doesn't involve long-lived radioactive waste, has a virtually inexhaustible fuel supply, is far safer and yet significantly more powerful.

CHAPTER 2

FUSION POWER

Einstein's famous E=MC2 equation significantly opened the door to understanding the workings of the stars and our sun. Scientists soon realised that there was a much better way of creating energy.

The stars are nuclear reactors of a different kind. They use a process we call 'Nuclear Fusion'. This takes simple light, safe elements and burns them into heavier safe elements. Nature's binding energy known as 'The Strong Force' releases its power as these elements fuse together and change giving out enormous energy in the form of heat and light. Our Sun is a large, local nuclear fusion reactor. We use its power to sustain life, maintain our eco system, grow our crops, power our solar panels and drive our wind turbines from the pressures it creates. The Sun gives life to our world in the most spectacular way and for man to truly master energy, he must learn to emulate its process.

Scientists have been aware of this potential power from as early as the 1930s but when research really got underway after WW2, they were faced with the only problem they are not trained to overcome: a shortage of funding. By the 1960s there was still not such a high demand for clean energy as there is today and fusion was becoming renowned for being extremely complicated after what some people were calling a failure at an experimental fusion reactor in the UK called ZETA. World leaders were not motivated to invest billions of dollars into fusion, as they had the highly weaponisable fission. Often fusion projects simply ran out of funding and closed down.

Two major factors kept fusion research alive. One being the fact that its ultimate success would be one of the most significant scientific and technological achievements ever in the history of mankind; the other being that no matter what obstacles, complications and expenses were thrown in the way, it's power producing potential was so great as to warrant finding a way around them regardless of cost. In short, when it comes to fusion, the juice is always going to be worth the squeeze.

The difficulty was convincing governments to bankroll the research. Private corporations were not positioned for such an undertaking. Nuclear research was to be government business and governments are notoriously bureaucratic, slow beasts.

In 2025, after what will be over 100 years of fusion research, the largest fusion reactor ever built is due to be turned on for the first time. It will finally demonstrate real fusion power to the sceptical media and financial investors. Overnight, there will be a great public appetite to understand nuclear fusion and get a grasp of its potential and the massive positive impact it will have on our civilisation.

Fusion in the Sun

Fusion is all about trying to bring little particles close enough so they might fuse together. In our Sun, a process of hydrogen fusion occurs.

Hydrogen is the simplest of elements, its nucleus consists of a single proton. You may remember diagrams on school chalk boards of a nucleus made up of 1 proton with 1 electron seeming to be in its orbit. This is hydrogen. It is a noble gas and 75% of the universe is made up of it.

In the Sun, hydrogen is fused into heavier hydrogen and then into helium and the energy release is obvious to us on Earth. For hydrogen fusion to occur, two protons need to be brought closely together. Once they get close enough, they will sometimes fuse.

The Sun brings these protons together by way of its heat. High temperatures are required so that the protons are moving fast enough for them to overcome their electrostatic repulsion to each other.

On an atomic level, the distance between a nucleus and its closest electron is vast; if the electrons are there, the protons have no chance of getting anywhere near each other. So, if you want to bring two protons close together, you must first get rid of their electrons. This process is known as 'Ionisation'.

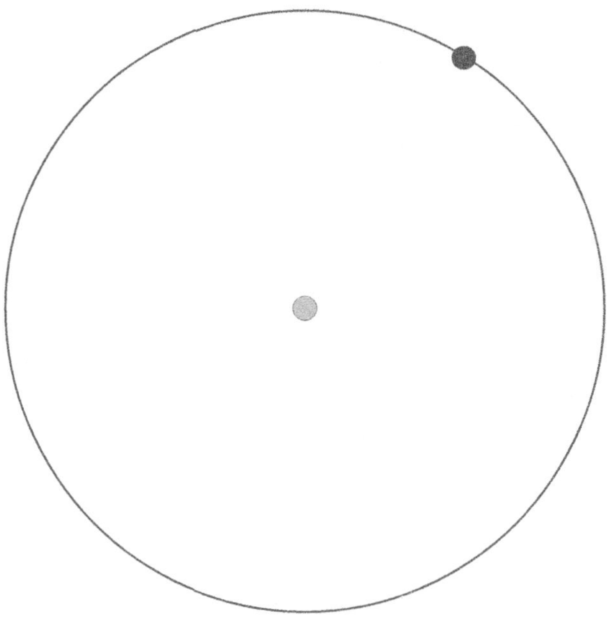

A hydrogen atom. The positively charged proton is shown at the centre and the equally negatively charged electron in its orbit. NB. electrons do not orbit in the nucleus in the same way the earth does the sun.

Fortunately, it is relatively easy to strip the electrons from an atom by creating what's known as a 'plasma'. Plasma is the 4th state of matter. We have solids, liquids, gasses and hotter still, plasma.

Temperature is a measure of the average kinetic energy of particles. In hot water, the particles are moving around much faster than in cold water. In a plasma, the particles start to move so fast that the electrons are torn off the nucleus leaving protons to move around on their own. These charged particles can now be called 'ions'.

In a plasma, the protons are now a great step closer towards their being able to fuse together but here enters the next obstacle: Protons are positively charged therefore just like magnets, they repel each other. So even though our plasma has allowed protons to whizz around in a much tighter proximity, they are always going to repel away from each other.

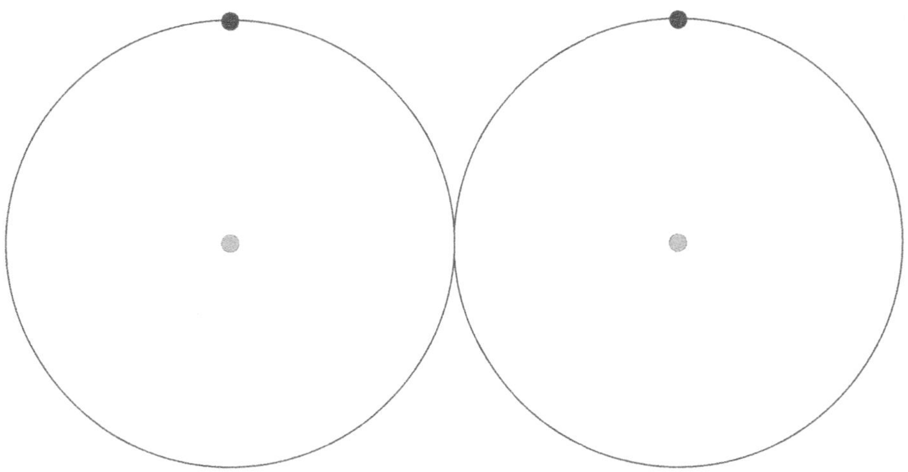

Two hydrogen atoms shown above. The electrons refuse to share space with each other. This keeps the protons at the centre from getting close to each other whilst the atom is in this state.

The plasma needs to be of extreme temperatures for the ions to be moving quickly enough to get sufficiently close for fusion to occur.

Here the hydrogen atoms have been 'ionised' and are free from their electrons. They can now get significantly closer but as they are both positively charged, they will still repel each other.

The Sun is a giant, hot ball of plasma. It fuses several hundred million tons of hydrogen into helium every second. Its massive size creates such pressure and temperature as to give the nuclei enough energy to overcome their electrostatic repulsion.

During the fusion process, when two ions get very close, the strong force locks them together and a funny thing happens: one of the protons spontaneously turns into a neutron. (A neutron is a particle of no charge). This newly created ion, consisting of a proton and a neutron, is an isotope of hydrogen known as a 'Deuteron'. This change from proton to neutron is courtesy of the 'Weak Nuclear Force'. Light is emitted as a product of this reaction which is why the Sun shines.

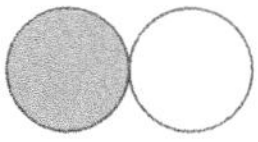

Here a proton has fused with another proton which has undergone a process called 'radioactive beta decay' and turned into a neutron. This proton and neutron is an isotope of hydrogen known as Deuterium which is found in abundance in sea water.

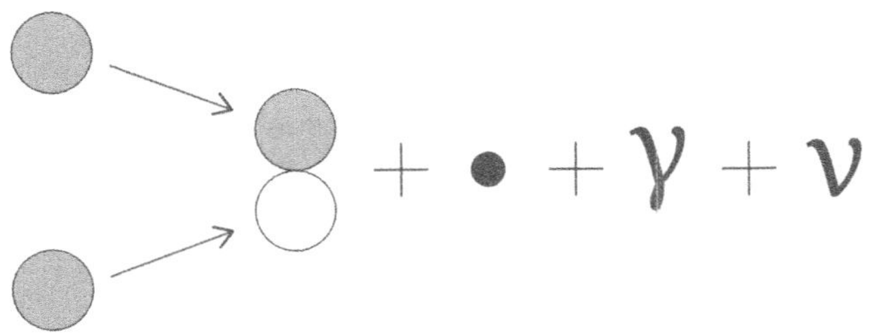

This process of change is achieved by a change in the protons quarks. A proton is made up of three 'quarks' two up quarks and one down. During beta decay, an up quark becomes a down through the 'weak interaction force'; A bi-product of this reaction is a positron which annihilates to create a particle of light called a photon.

A deuteron can then fuse again with another deuteron inside the Sun's burning plasma. The product of this Deuterium – Deuterium fusion (D-D Fusion) is helium and vast amounts of energy in the form of heat.

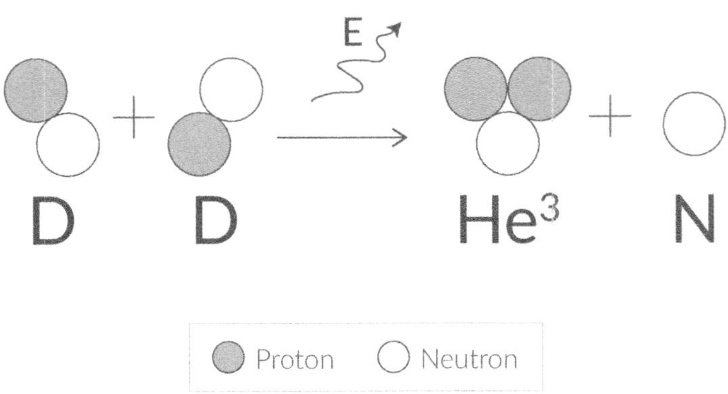

Here two deuterium ions have fused to make Helium 3 and a Neutron. This is one of the reactions going on inside the Sun. The neutron shoots out at high speed and its kinetic energy creates the additional heat.

We would like to emulate this deuterium-deuterium fusion reaction on earth for the reason that it produces a million times more energy than a chemical or combustion reaction.

There is a reason scientists have been trying for 100 years to achieve this and there are quite a few problems to overcome in order to create the conditions needed for this to happen on Earth. The fairly obvious one is to emulate the temperatures of the Sun in a laboratory. The Sun has gravity on its side. Its gravity is 28 times stronger on its surface than that of earth's due to its size (the Sun is about 1 million times bigger than Earth). This contributes to its immense heat and particle fusing potential. We clearly can't build something that size on Earth but the problem has inspired scientists to innovate some ingenious and fascinating devices.

Fusion in a Machine

To get fusion to happen in a machine requires that we have a plasma of temperatures no less than that at the core of the Sun (27 million degrees Fahrenheit) and far hotter still to compensate for the absence of size.

Currently, power stations use an energy source to create mechanical power, such as turning a turbine. By converting this mechanical power into electricity, national grids are supplied. Energy sources vary, but include fossil fuels such as coal, nuclear fission, and renewables such as water or wind.

Nuclear fusion has not yet been achieved to the extent that it produces a net energy gain, but scientists are determined that it will. The path to achieving this centres around what type of reactor to use.

The most popular reactor is known as a 'Tokamak'. It is the model for the reactor at ITER and most scientists believe it is the most viable option for producing the kind of sustainable nuclear fusion that could power a national grid. The name 'tokamak' is derived from Russian and literally translates to 'toroidal chamber with magnetic coils.' Essentially, the tokamak reactor is in the shape of a ring doughnut.

It poses fewer (though still significant: ITER's tokamak will require one million components) engineering challenges when compared with some other fusion reactors. Inside the chamber, plasma is held and controlled by enormous magnetic coils, which stop the hot plasma from coming into contact with the reactor's walls. The fusion reaction happens continuously, not in burst, and so is a stable option for any future power plants.

The simple science behind how a tokamak might produce electricity relies on the heat generated by a nuclear fusion reaction. The reaction inside a tokamak produces enormous quantities of heat, which is then used to produce steam to power turbines, and thus, electricity.

Two principal alternatives dominate the landscape in the UK and US: inertial fusion energy (IFE) and extreme compression. In IFE, lasers are fired at a pellet of fuel until its atoms fuse. Extreme compression, on the other hand, uses pistons simultaneously to fire a shock wave of energy into the plasma, compressing it to the point of fusion. This has only been possible to achieve since developments in computing have allowed complete synchronicity to be established.

The basic working principles of these devices are as follows:

Firstly, we need to build a high vacuum chamber in the shape of a doughnut, known as a 'torus'.

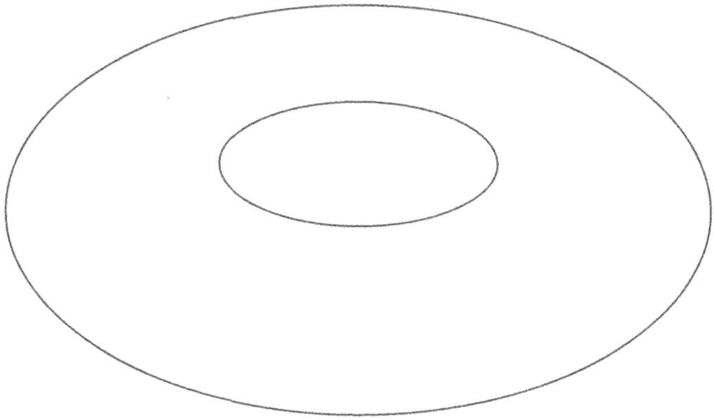

As plasma is made up of charged particles, we are actually able to manipulate and control it using electromagnetic fields. Powerful magnets are placed around our vacuum vessel as shown below.

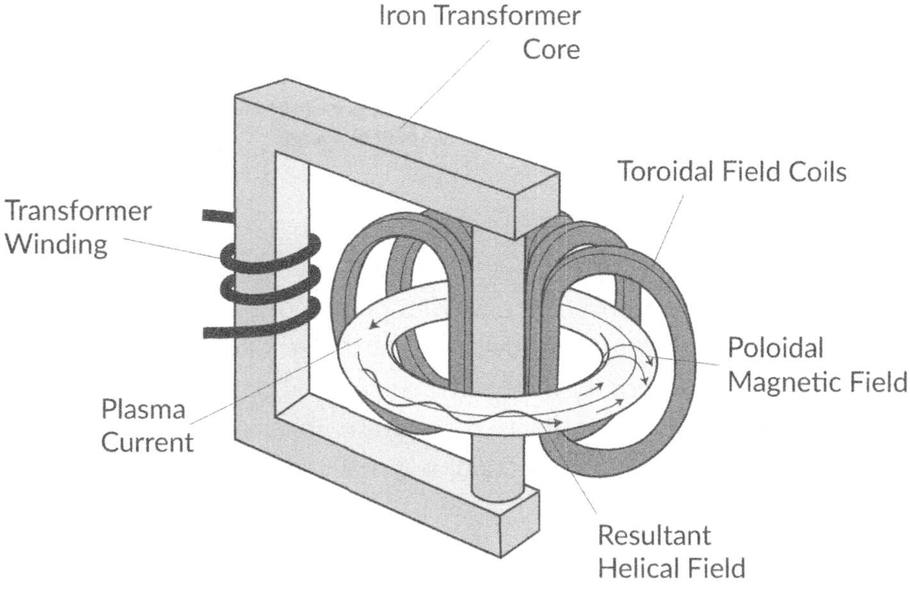

The vacuum chamber is filled with deuterium and tritium gas, radio frequency heating is then used to ionise the plasma. The plasma can be held in place in the vessel away from the walls, using the electromagnetic fields created by the magnets wrapped around the torus.

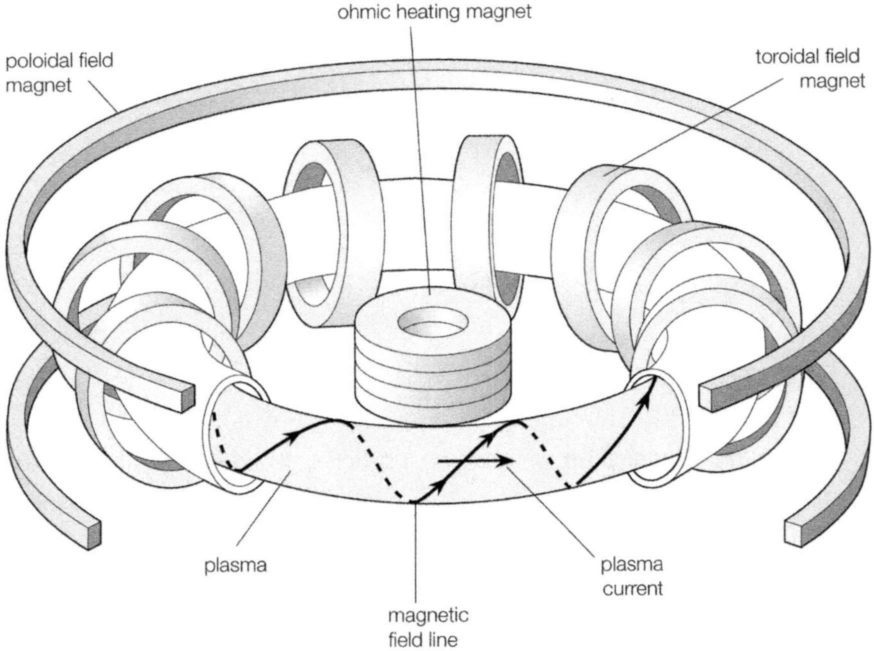

DD fusion, the reaction produces a high-energy neutron and helium 3. This reaction yields 3.27 MeV of energy.

In this device, we can have a controlled plasma of high temperatures up to 300 million degrees, confined and controlled safely on Earth. The plasma is extremely hot but should the current fail, the plasma will stop and revert back into its pre-ionised, gas state. Tokamaks are extremely safe as their damage or destruction only results in the plasma turning back to its original gas.

We can choose what isotopes of hydrogen to put into a reactor. Fortunately for us, we already have an abundance of deuterium on Earth in our oceans. We therefore do not need to perform hydrogen -hydrogen fusion to create deuterium as we can extract it from sea water.

We could fuse deuterium with deuterium to create helium, (D-D Fusion) but there are other, more readily fusible hydrogen isotopes available to us.

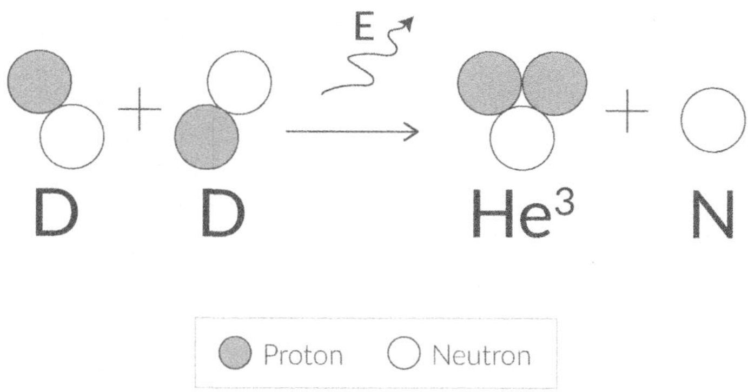

DD fusion, the reaction produces a high-energy neutron and helium 3. This reaction yields 3.27 MeV of energy.

'Tritium 'which consists of one proton and two neutrons, is a radioactive isotope of hydrogen. The plasma temperatures required for deuterium to fuse with tritium **(D-T Fusion)** are less than those required for deuterium to fuse with deuterium **(D-D Fusion)** and more energy is produced as a result.

If we bring deuterium and tritium ions together in our Tokamak, the result will be as follows:

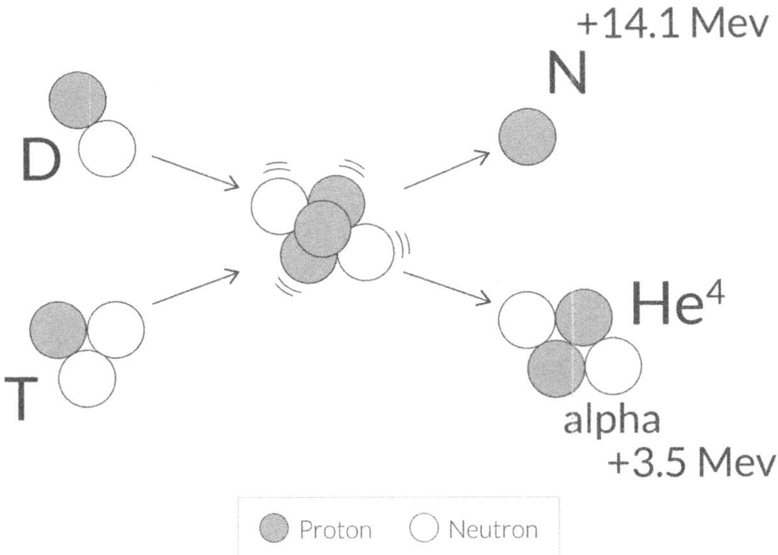

DT fusion, the reaction produces a high-energy neutron and helium 4. This reaction yields 17.6 MeV of energy.

The product of the D-T fusion reaction is an alpha particle (which is Helium 4) and a high speed neutron. The kinetic energy of the neutron is responsible for 80% of the energy released in this reaction.

As the neutron is a neutral particle of no charge, it will not feel the electromagnetic confinement of the tokamak. It is free to escape the confinement area of the hot plasma and shoot out at very high speeds in any direction. In a D-T plasma, the neutrons slam into the reactor walls depositing energy as heat. We can use this heat to drive a steam turbine and generate electricity.

To understand how this device succeeded brilliantly in achieving the necessary conditions for fusion on Earth, and yet why fusion isn't already powering our homes and society, we must first take a trip back into the past to when physicists first comprehended the magnificent prospects of fusion and started creating the first devices to confine a burning plasma.

CHAPTER 3

CONFINING PLASMA

The Pinch

Since fusion development began in the 1930s, innovative devices for trapping plasma have not been in short supply. In 1947, Peter Thornemann, an Australian physicist working in the UK commenced construction of a glass ring tube designed to hold plasma which became known as a 'torus'. Thornemann discovered how to initiate a plasma using an electromagnetic field and get the plasma current to flow around the torus. He used the principal of the 'pinch effect' which had been discovered when lightning struck a metal tube on a rooftop. The resulting electromagnetic forces pinched the tube and crushed it inwards.

This tube has been 'pinched' by the electromagnetic forces created as the result of a lightning strike.

Thornemann used this effect to pinch the plasma in the centre of the torus ring. Electromagnetic induction would push the current around the glass in a similar way that electric current can be driven down a copper wire using magnetic fields.

Although the pinch device was not capable of getting plasma near to the conditions necessary for fusion to occur, Thornemann continued working on the idea. His pinch had kicked off the study of plasma behaviour under magnetic confinement in a big way. He experimented with mercury gas in his pinch device and learned to arrange electromagnets in a way to centralise the plasma in the vessel. He then went on to build a larger copper device but ultimately knew that a much larger, more powerful tours was needed.

ZETA

In 1957 the ZETA (Zero Energy Thermonuclear Assembly) reactor in Harwell, UK went into operation. It was the largest device in the world at the time and it incorporated into its design many aspects of Thornemann's pinch.

It showed an improvement in stability and plasma time of up to a few milliseconds. To the delight of its creators, neutrons were recorded being emitted from the reactor which scientists hoped evidenced that lots of nuclear fusion reactions were taking place inside the plasma. Sadly, for the ZETA team these neutrons were proven not to be thermonuclear. The temperatures were not high enough for a meaningful fusion rate as ZETA was only able to reach up to 5 million degrees.

A machine capable of temperatures of at least 100 million degrees was clearly needed for significant amount of fusion reactions to occur. A small number of nuclear fusion reactions did take place in the ZETA plasma but ZETA 2 was not subsequently funded. Instead it was decided that scientists first needed to better understand the theory of plasma behaviour under confinement before another major reactor was to be funded.

The ZETA reactor in Harwell, UK in 1957

Bohm Diffusion

There is an expression in the fusion community that goes; 'Fusion is easy, plasma physics is hard.' In fact, it was the theoretical complexity of plasma physics that ultimately brought about the demise of ZETA and its planned successor.

Keeping the plasma going was the major issue in these early reactors. The plasma would become unstable and fail. One of the effects that physicists discovered was a variation of classical diffusion in the plasma called 'Bohm Diffusion' after David Bohm.

In the Plasma, the ions 'corkscrew' around the magnetic field lines.

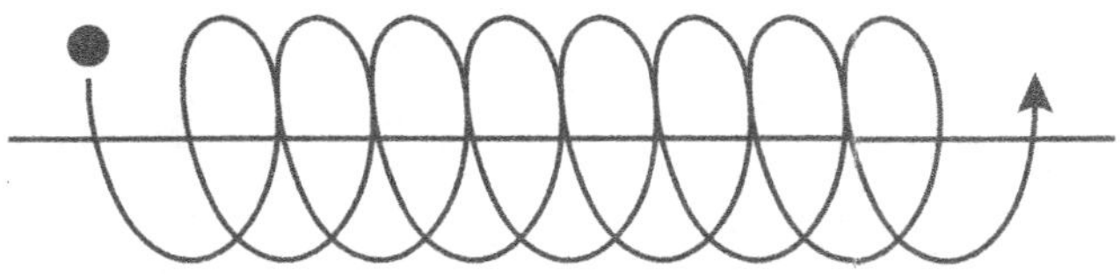

One of the problems was, in the torus shaped reactors, the ions were drifting towards the outer field lines and ultimately being transferred into the walls of the reactor. This was transferring heat from the plasma into the reactor walls.

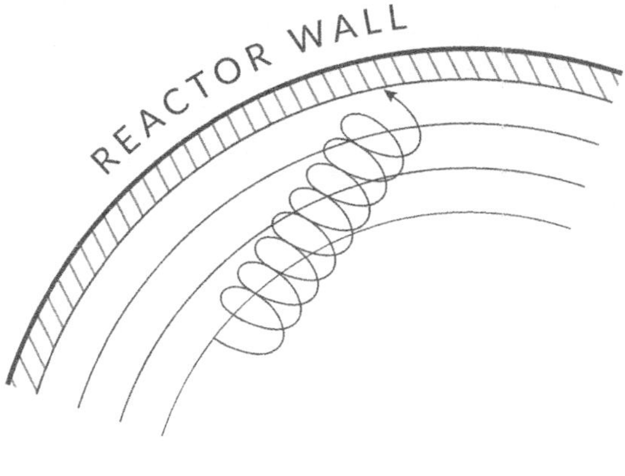

When scientists were concentrating on creating the highest possible plasma temperatures to maximise the fusion rate, this was a very unhelpful side effect to be experiencing.

The Magnetic Mirror

The Ideal solution to the problem to the diffusion seen in torus reactors would be to get rid of the shape altogether and just have a straight tube reactor. If your reactor is straight, you get a nice linear plasma flow and less diffusion into the reactor walls. But you now have the question of what do you do at the ends?

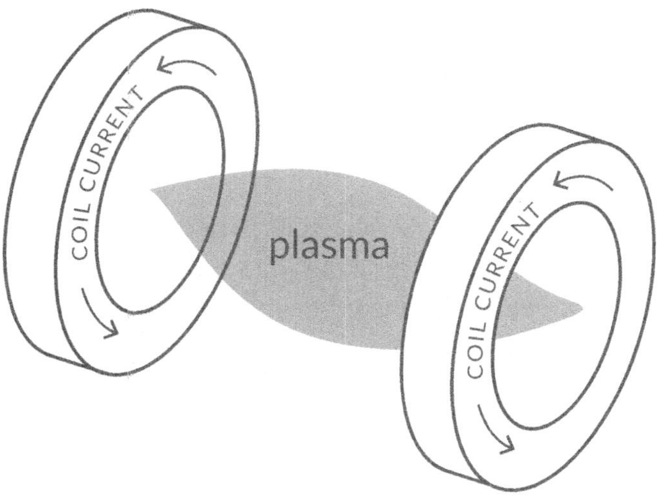

The solution was to have several magnets at each end that worked like 'mirrors' to drive the ions back into the plasma. The success of this device has been limited however as you need complex magnet arrangements on each end to stop the ions leaking out. There is a saying about magnetic mirror devices that compares these multiple magnets to the ring lines in a tree trunk, claiming you can tell how old such a device is by how many magnets it has at its ends.

The Stellarator

One elegant solution to the problem was conceptualised by Lyman Spitzer in 1951 whilst pondering this issue in the Alps, he came up with a reactor shaped like the figure of an 8.

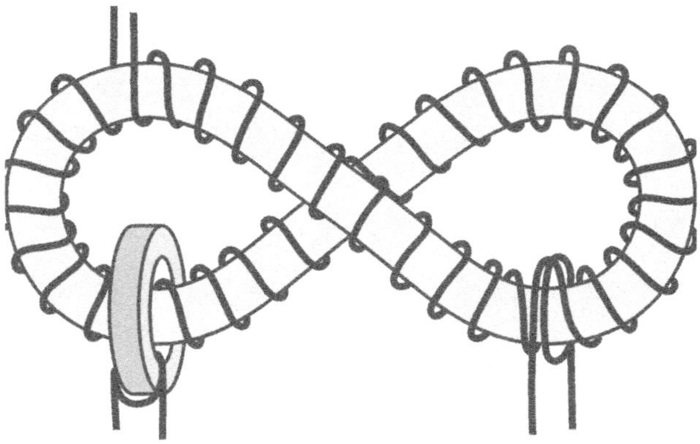

Comparable to a remote-control car going around a race track, if the car is on a circular track and the operator runs it too fast, the car will be sent flying off the track much like the particles in the torus reactor. In the 8 shape, the particles are driven back inwards at the opposing corners.

The stellerator is deservedly much acclaimed and today the Wendelstein stellarator in Germany stands testament to its ability, recently achieving first plasma. The downside to this design is that it has become an awesomely complex device costing billions of Euros to construct and made possible only by the design capabilities of modern supercomputing. This complexity may ultimately be its downfall as the world will need simple, cost-effective, repeatable reactors that can be easily maintained.

CHAPTER 4

THE TOKAMAK

I have described one of the challenges in the quest for nuclear fusion as recreating in the lab the conditions at the centre of the Sun. This in itself sounds like a herculean task. However, scientists have made considerable strides in this area, finding a number of ways around some of the more complicated aspects.

For example, in solar fusion, the process is slow, requiring two protons and two neutrons from the nuclei of hydrogen atoms. It takes, on average, approximately one billion years to complete this reaction. Physicists have established that there is a much quicker way. Most nuclear fusion projects now work with deuterium and tritium, hydrogen isotopes which carry one (deuterium) or two (tritium) neutrons. Tzhey react with each other much more quickly, producing helium and one extra neutron. It's more efficient and both are relatively abundant: deuterium can be extracted from water, and tritium can be created synthetically from lithium, which is widely available. This discovery has had implications for the design and engineering of reactors, as it reduces the need for reactors to be able to cope with the large levels of antimatter produced in solar nuclear fusion.

In designing reactors that are capable of heating plasma to temperatures higher than those of the stars, and which are then able to safely contain the reaction, a number of innovations have been made: magnets with uses in healthcare, and software capable of simulating plasma burning at extreme temperatures. So the search for nuclear fusion energy is already yielding positive results, even if they happen to be a by-product of the process. Achieving energy-gain fusion is the end goal, but the scientific and technological discoveries we make along the way will be of benefit to people for generations to come. Furthermore, established areas have come into their own; 3D printing, for example, has been of crucial importance when designing and building reactors, especially in smaller private enterprises.

Laboratories across the globe have achieved a number of world firsts. In 1991, the JET experiment at the Culham Centre for Fusion Energy produced the first controlled release of deuterium-tritium power. The same reactor also holds the current world record for fusion reaction output, achieving 16 MW in 1997. More recently, scientists in China have created hotter and longer fusion reactions than ever before. In 2017, the Experimental Advanced Superconducting Tokamak (EAST) sustained fusion conditions for over 100 seconds, which has never before been achieved. And just last year, physicists in the same laboratory produced plasma temperatures of 100,000,000°C - six times the heat at the centre of the Sun.

JET

In the 1950s the 'Tokamak' was invented by Soviet physicists Igor Tamm and Andrei Sakharov. In 1968, Russia achieved stunning temperatures in their Tokamaks and announced results that caught the attention of the world. This triggered a major move from torus reactors to Tokamaks. The key difference between Tokamaks and the existing pinch torus designs was the pinch device was designed to compress the plasma very rapidly whereas the Tokamak was designed to hold the plasma in place.
Perhaps one of the most vital influences of this announcement was over Paul Rebut who was working on a large toroidal device for Europe in 1971. Rebut promptly switched his attention from torus pinch designs to Tokamaks. He led the research for a torus reactor called JET - (The Joint European Torus). JET became a tokamak in design but was named torus, as it was felt, that it sounded less Russian.

A look inside JET which is based at the Culham Centre for Fusion Energy in the UK.

EUROFUSION

JET nearly never came about after a decade of debate over its proposed location. As a largely EU funded project, there was strong opposition to it being constructed on British soil. There was also lengthy debate over countries' respective share of funding which in turn became location sensitive; and then further arguments over its dimensions. JET was not actually given the green light until 1977 when it was finally agreed that JET would be built at the UK's Atomic Energy Authority in Culham near Oxford.

JET finally achieved first plasma in 1983 and it was officially opened by Queen Elizabeth II in 1984. The reactor was the first of its kind to feature a 'D' Shape vessel as opposed to being circular. The reasons for this were partly financial and partly as the magnetic field strength is stronger closer to the central solenoid magnet. It was also the largest tokamak of its kind as it was generally felt the operating conditions would improve with size. The plasma would carry currents of up to 5 Mega Amperes (MA) with a plasma volume of 90 cubic meters.

Neutral Beam Injection

To increase plasma temperatures in Tokamaks, several methods have been developed to further heat plasma. Radio Frequency Heating and Neutral Beam Injection are two ways of boosting plasma temperatures.

The method of introducing neutral beam injection heating without destabilising the plasma was discovered in 1982 by Fritz Wagner at the Max Planck Institute for Plasma Physics in Germany. The Wagner method of using a higher density of particles became known as 'H mode' or high density mode.

After successful demonstration of the H mode effect at General Atomics in San Diego in 1986, JET looked to incorporate Wagner's method as they had been experiencing poor confinement from their own neutron beam injection efforts.

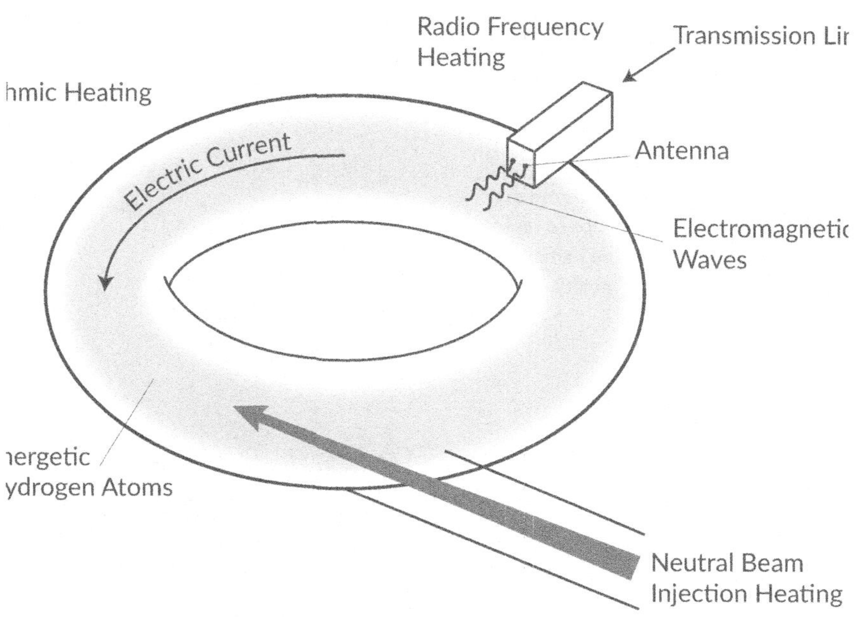

Divertors

An important requirement for retaining plasma temperature and confinement is keeping it as concentrated with your deuterium and tritium ions as possible.

As ions diffuse and hit the reactor walls, they knock loose small parts of the metal and these may contaminate the plasma. The tiny fragments of metal retain their electrons in the plasma and allow heat to escape in the form of photons as the metal itself starts to ionise.

A divertor is a device fitted into the bottom or top of the D cross section in a modern Tokamak that shaves off the outer layer of the plasma removing potentially contaminated parts and allowing gas and heat to escape in a controlled way. In 1986, H Mode neutral beam heating was thought only to be achievable in Tokamaks fitted with divertor. This led to JET incorporating a divertor at the bottom of the Vacuum Vessel.

The modern MAST (Mega Ampere Spherical Tokamak) reactor which is also based at Culham, is fitted with two of these divertors at the bottom and top of the vessel. Divertors have proven to be an extremely important innovation in modern tokamak design.

The divertor is placed at the bottom of this D shaped tokamak. It helps to remove impurities, act as a controlled exhaust system and assists with retained stability.

EUROFUSION

PULSAR FUSION

Supershots

In 1985 at the TFTR (Tokamak Fusion Test Reactor) in Princeton, New Jersey, scientists found the carbon coated walls of the reactor had been absorbing deuterium and other ions from the plasma. Upon reheating, these ions were re-emerging in the new plasma. This was altering the expected plasma densities of future experiments. It was the carbon coating that was absorbing these ions and then spitting them back out. In 1986, Jim Strachan decided to attempt to 'wash' the carbon walls by running multiple helium plasma shots. Helium is non-reactive so Strachan reasoned this would help to clear out any remaining ions from the carbon walls.

Following his helium 'washing' process, subsequent deuterium shots were so improved in temperature, density and confinement time they became known as 'Supershots' and led to plasma temperatures of up to 400 million degrees being achieved in the 1990s.

The 'Supershots' story to me serves as an important reminder that significant discoveries are sometimes made by chance once you get practical. This would not have been discovered on a whiteboard by theoretical physicists. Sometimes you just need to build machines and start using them to make discoveries like this.

Turbulence

Perhaps one of the most significant factors that prevented smaller Tokamaks from achieving long confinement times is a plasma disruption factor of toroidal rotation caused by 'MHD turbulence' (Magneto Hydro Dynamics). This is arguably the main reason for the size of the World's largest experimental Tokamak, ITER (International Thermonuclear Experimental Reactor). Like water running from a tap, at slow speeds there are no problems but as you turn up the pressure, it starts to get more violent.

The larger the vessel, the less of a problem this turbulence is. The shape of the vessel can also play a key part in reducing this effect. Due to limited funding, not as many tokamak vessel shapes have been practically experimented with as researchers would like. The 'D' shape being the most popular today is one of only a few shape variations actually incorporated in real world large Tokamaks. Professor Steve Cowley, former CEO of the UK's Atomic Energy Authority gave a lecture at Imperial College London titled 'Shape Matters – Building a better bottle' which highlighted this lack of experimentation.

At the time of writing, a method of totally preventing turbulence in Tokamaks has not yet been devised.

Some scientists believe that using stronger, more powerful superconducting magnets will help force the plasma into submission but there is a point where the power of the magnets would be overbearing on the structural capabilities of the tokamak. Others feel we need to further understand particle behaviour so we are able to build a better shaped reactor rather than force it to perform using the current shapes. Modern super computer simulations have already made significant contributions to modern tokamak design. There is still much work and practical testing to be done in harnessing the sciences of MHD and Gyrokinetics.

Were it not for the effect of turbulence, Tokamaks could theoretically be produced the size of a football.

Spherical Torus

One change in tokamak design that has reduced the turbulence effect is the reshaping of the vessel. This was an innovation first proposed in 1984 by Martin Peng of the Oak Ridge National Laboratory USA. The Spherical Torus (ST) is the move from the conventional doughnut shape to a cored apple.

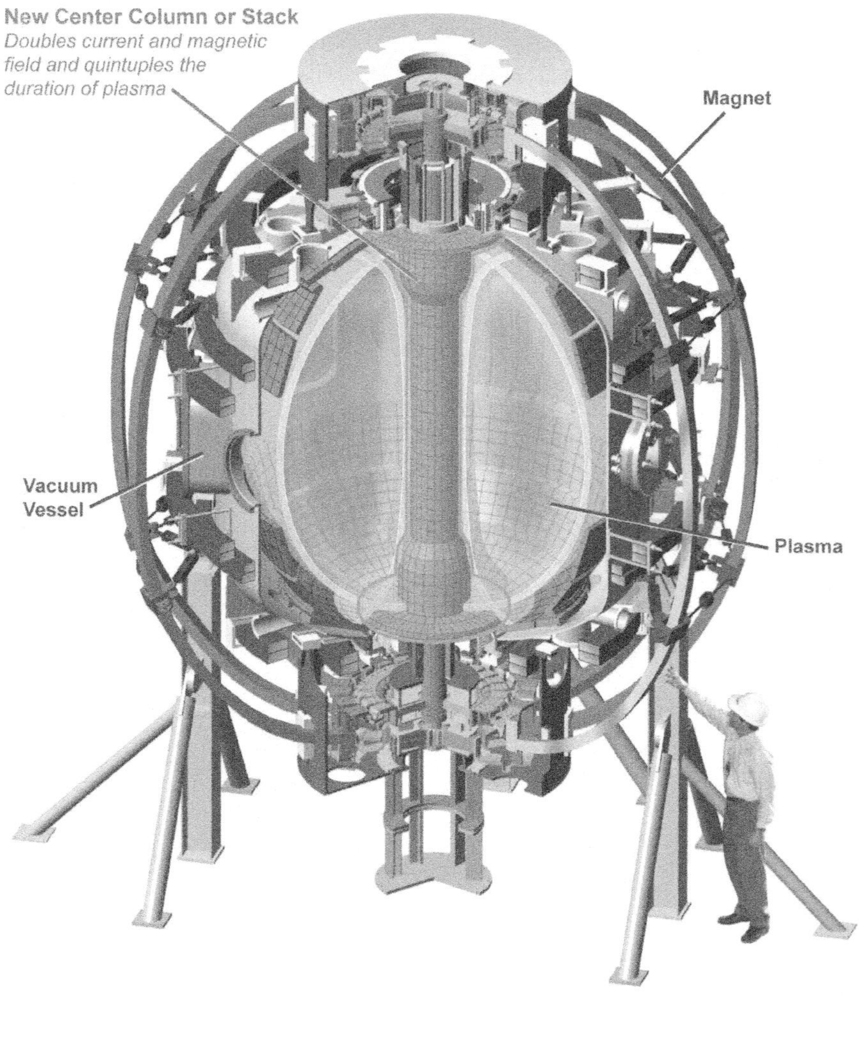

ST designs have improved confinement and pressure limits. The shape reduces available space for the powerful central solenoid magnets so new ways to fit them in have been devised.

The advantage of the ST design is rather than wrapping individual magnets around the chamber, you can have a main central conductor and link half ring magnets to it. This reduces the electromagnet mass of the design and means the ST is able to achieve better confinement with less magnetic force.

Fusion is achieved

On 9 November 1991, Scientists at JET conducted a tritium experiment which achieved the world's first controlled release of fusion power. In 1997, 16MW of power was extracted from the JET fusion reactor.

This result is due to be exceeded very soon by a larger reactor called ITER which aims to extract 500MW from fusion. Sadly, there will have been a 30-year waiting period for its construction. This is comparable to apple successfully creating the iPhone and then arguing internally for 30 years over where their first store should be built. It highlights the operational and motivational difference between governments and private corporations.

CHAPTER 5

THE MIGHTY ITER

Most people will be familiar with the fable of the hare and the tortoise: two animals enter a race, both keen to cross the finish line before the other. The hare - sprightly, confident and the overwhelming favourite to triumph - is beaten by the slow and steady tortoise, who doesn't rest until he has won.

And so, it is that two opposing sectors have found themselves competing in the quest for nuclear fusion energy. International, government-funded projects such as the International Thermonuclear Reactor (ITER) in France, which is backed by 35 countries across the globe, is the public sector mega-project. In its fourth decade, over budget and way behind schedule, it is finding itself in an increasingly crowded marketplace, as the private sector populates the field with numerous ventures, funded by the likes of Amazon's Jeff Bezos and the late Paul Allen of Microsoft. Innovative, agile and ambitious, these startups are focusing on smaller experiments, taking advantage of the very latest in technology.

Public or private sector: who will be first to achieve the ambition that has eluded generations of scientists, to realise nuclear fusion energy? Tim Luce, Chief Scientist at ITER, puts it simply: 'we're the tortoise'.

By the 1980s tokamak reactors had clearly demonstrated that fusion could be made to work. The successful demonstrations at JET and NSTX in Princeton PPPL had achieved the conditions necessary for net energy gain fusion. This means that had there been a tritium-deuterium fuel in the reactor, they would have achieved a net energy gain.

To calculate a net energy from a fusion reactor one must record how much energy is created minus the amount of energy used to run the reactor.

Fusion Reactor performance is measured by the value of 'Q'.

Q=1 being break even, the fine point at which the reactor is producing exactly the amount of energy it requires to run itself and anything over being an energy gain.

ITER aims to produce Q=10. This means it aims to produce 10 times the energy that is required to run the fusion reactor. Many scientists I have spoken to working on ITER believe it may very well achieve Q=15 quite easily.

Delays

In 1986 funding for ITER (International Thermonuclear Experimental Reactor) the world's largest experimental reactor, was approved. ITER also translates as 'The Way' in Latin and is now hotly considered to be 'the only way' by many scientists. Weirdly pronounced 'eater' after being located in France, it is now the pride of a multi-government fusion effort. It is funded and run by seven member nations:

China

The European Union *(if you consider that one entity)*

India

Japan

Russia

South Korea

The USA

The EU is stumping up 45% of the bill with the other six countries contributing 9% each. It is the perfect storm for bureaucratic mayhem.

From the word go, it took the leading fusion physicists from the 7 respective countries an achingly long 11 years to complete the design.

Subsequently, the first mandate given to them was that a site must be chosen for the reactor by 1998 and construction should be completed by 2008. The budget was set at the equivalent of 6 billion Euros.

The government representatives from the respective nations began arguing over the site location and no progress was made on the world's most important, clean energy test reactor for almost another decade whilst this went on.

In 2006, (8 years after their deadline) the final site was eventually decided on in Cadarache, Provence, France. The frustration of the physicists who had been committed to working on ITER must have been close to unbearable with total years lost to politics amounting to well over a decade by this point. Even in the 1990s, private companies were not in the position to step in and build their own reactors. Nuclear was exclusively a government game.

ITER is now not due to run its first plasma until 2025 and its cost has now risen to 13 billion Euros.

Size

For governments to agree this large funding, they understandably wanted to be absolutely certain that the reactor would work. If ITER can demonstrate controlled Q+10 fusion, then a planned demonstration fusion power station will be financed out of its success, called DEMO.

Fusion had a reputation for being complicated. Scientists had to convince the government funding powers that ITER would work. They had to navigate around the problem of the not fully understood turbulence disruptions. One way to do this was size. In short, scientists found that the larger the reactor, the less of a problem these plasma disruptions were proving to be. So back in the 1990s, size was the only way scientists could guarantee ITER's success.

ITER

The ITER constructions project - France, as seen from above in 2016

After much deliberation, ITER finally came in at 23,000 tonnes which is an immense amount of metal work considering the Eiffel tower is only 7,300 tonnes. It will be 60 meters high with an 840 cubic meter vessel built on a construction site of a whopping 42 Hectares.

Essentially, ITER was made so large as a direct result of a lack of understanding of the control and behaviour of plasma. Originally based on a 1990s design and without today's supercomputing powers, size was simply the only sure solution. The downside to size is great cost and construction time. A total 20 years minimum build time on what is now no longer a science experiment but a construction demonstration.

Since 2006, the cost of supercomputing has crashed and simulations are showing us that reactors do not need to be the size of mountains or cost $14 billion. Modern gyrokinetic simulations are demonstrating realistic tokamak parameters a fraction of the size ITER.

It is too late for the ITER team to incorporate the innovations regularly coming out from the reactors at Culham, Princeton and the findings from universities. Incorporating any suggested design improvements now would increase the cost and time of completing the project. ITER just needs to finish construction, demonstrate and wake the world up to the potential of fusion power.

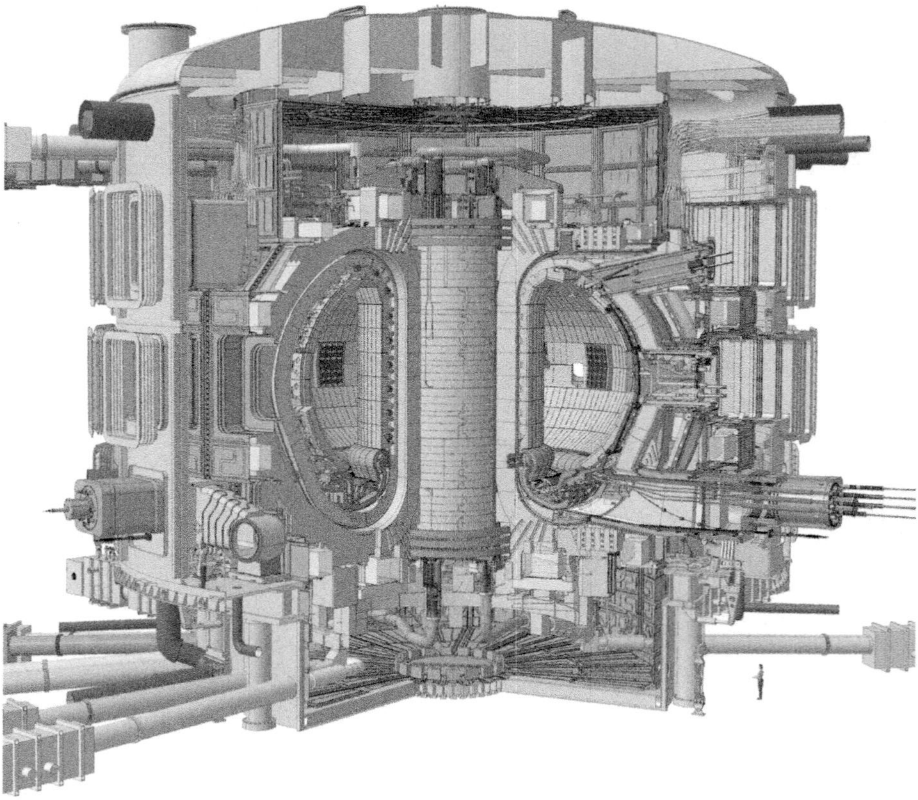

ITER

On the right-hand bottom corner you might be able to spot a human silhouette for scale.

The great thing about ITER is that when it does turn on, it will finally bring nuclear fusion out of the land of the mythical and into the forefront of humanity's interest, where it should be. The world is covered in scars from the petty scratchings of man, but ITER is something we should be truly proud of.

Bernard Bigot, Director General of ITER, sees this global project as the silver bullet to our energy problems: 'As a scientist I have been looking for this technology for several decades... If we succeed it will be a real breakthrough for the world's energy, not only in this century but for millions of years.'

We have discussed that the project is delayed and over-budget and that ITER is backed by the governments of 35 countries (the European Union, China, Korea, Japan, Russia, the United States and India), who share in the costs and the results of the project. Of this, the EU is the largest contributor, putting up 45.6 per cent of construction costs, while the remaining six nations chip in an equal share of 9.1 per cent each. Contributions come in the form of components and infrastructure in the main, with ITER member states manufacturing parts, building materials and systems that are then shipped to the project's site in the South of France. Countries top up these contributions with 10 per cent of their stake coming in the form of a monetary donation.

Born out of a sense of idealism, this mega-experiment has been designed to demonstrate that cooperation will win out over competition. With diplomacy at its heart, ITER is not just about dividing up the costs: it plans to show the world that through international collaboration, no country needs to be left behind in the search for nuclear fusion energy. ITER will share the knowledge that its research will produce.

And yet it has been international cooperation that has been one of the central risks to ITER; with political tensions almost throwing the entire project into orbit. In 2015, new management was brought in, in the form of Bernard Bigot. His priorities: efficiency and milestones. Over the past four years, changes have certainly been made, and public perception of the project is turning around. Bigot is cautiously optimistic: 'Everyone is now complying with new best standards for project management. The atmosphere of the project has completely changed too. People believe that fusion is on track. Before it was almost a dream. The next few years will be critical to reinforce the trust we have in our capacity to deliver.'

But challenges remain, particularly in the current political context. With a rise in protectionism across the West and the impact of a potential Brexit on the horizon, ITER is navigating the developing faultlines amongst the international community.

Nonetheless, the organisation's goal is clear: 'ITER will be the first fusion device to produce net energy', is the claim made on its website, and at present, it represents the safest bet. Once built, ITER's reactor will be the world's largest tokamak, capable of creating and sustaining fusion reactions in a way that, to date, no-one has achieved. The project will not produce electricity from the energy created in its reactor, but will pave the way for the fusion power stations of the future: its successor, DEMO, which will produce electricity, is already in the planning.

How will it do it? The key is in the design of its tokamak reactor. When it comes to the engineering side of the equation, size matters: the larger the reactor, the more plasma it can contain. And more plasma means more fusion reactions, which in turn mean more energy. The ITER tokamak will have ten times the plasma volume of any machine in existence today, and as such has been designed to produce 500MW of fusion power - more than 16 times JET's current 1997 record. Scientists hope that the ITER plasma will eventually remain stable enough to demonstrate how a potential fusion power plant could work in the future.

But - and this is crucial - will ITER succeed? Can a project which, to date, has been characterised by delays, the haemorrhaging of money and, at times, questionable management, turn things around, harness its talents and resources, and deliver its grand plan? Its proponents argue that ITER is a bastion of collaboration, paving the way (originally International Thermonuclear Experimental Reactor, it has now been retrofitted with a more inspiring Latin name) for fusion electricity. Cynics, however, claim that ITER is a quagmire, sucking in time and money: the first experiments have already been pushed back from 2016 to 2025, and estimated costs now reach $20 billion, which is at least twice the original budget.

This is where the controversy lies, and delays at ITER have created the space for others to join the fray. As the private sector builds commercial reactors looking to create fusion energy well before 2025, scientists at ITER are under pressure to make sure they don't become an also-ran.

CHAPTER 6

TURNING FUSION INTO A REALITY

In a small town near Oxford in the UK is Culham and the CCFE (Culham Centre for Fusion Energy). It is the home to two very important Tokamaks, JET and MAST (formerly START).

Founded in 1983, it has played a vital role in the global quest for nuclear fusion. Culham's Joint European Torus (JET) project and Mega Amp Spherical Tokamak (MAST) are a key focus of research as part of the preparations for ITER. At present, some 850 scientists and engineers are based at the 80-hectare site. CCFE receives funding from both the UK government, in the form of the Engineering and Physical Sciences Research Council, and from the EU under the EURATOM treaty (which the UK will be leaving as part of the implications of Brexit). It has recently been guaranteed further EU funding to 2020. Beyond that, its future remains uncertain.

CCFE hosts two of the most significant nuclear fusion projects in the world: JET and MAST. JET is currently the largest magnetic fusion experiment on the planet, though will be superseded when ITER is built. Currently the primary focus of nuclear fusion experiments in Europe, JET is the biggest and most powerful tokamak on Earth. It runs on deuterium-tritium fuel, which ITER will also use, and which many scientists believe will be the basis of commercial fusion power. JET's achievements to date have convinced physicists that a deuterium-tritium-fuelled tokamak is the most viable option for ITER's nuclear fusion reactor.

Some of the key milestones achieved at JET have included the first ever controlled release of deuterium-tritium fusion power in 1991. JET also holds the current world record for fusion reaction output, achieving 16 MW in 1997. When it comes to plasma research, reactor design and engineering, JET is a key facility to determine the shaping of ITER.

The other major project at CCFE is MAST (Mega Amp Spherical Tokamak): one of the two most important spherical tokamaks in the world. MAST scientists work in partnership with physicists at the NSTX-U experiment at Princeton in the US, testing the potential of spherical tokamaks (as opposed to toroidal ones, such as JET and the proposed reactor at ITER).

MAST is significant because it is one of the only publicly-funded spherical tokamak projects in the West, with toroidal models typically preferred by government-backed initiatives. The science behind ITER is tested in a different environment, allowing scientists to establish whether spherical tokamaks could one day provide the basis for a fusion power station.

PULSAR FUSION

MAST has produced over 30,000 fusion reactions, producing huge volumes of data and analytics on the best ways to create a plasma and also to keep it stable. A major, £30 million upgrade is now being implemented, to make MAST more powerful and more flexible. MAST will soon be able to create hotter conditions, have longer pulse lengths, and control and contain plasmas with higher temperatures and longer pulses. The upgrade will create a pathway towards the design of a future nuclear fusion power plant, as scientists will be able to observe and analyse 'steady state' plasmas. These plasmas will form the basis of the kind that will be used to generate electricity, running for hours or days - not just the seconds of today's plasmas.

Explaining the reasoning behind the upgrade, CCFE's CEO, Ian Chapman recently told the BBC: 'In the middle of these reactors you have a fuel that is 10 times hotter than the centre of the Sun. You have to have a stable and safe way to get the exhaust heat escaping from the plasma out of the reactor without melting any parts of it.' Not only that, but MAST's upgrade will allow scientists to find materials capable of withstanding the conditions inside a fusion reactor: 'A fusion reactor will be the most intense source of neutrons on Earth,' argues Chapman. 'We need to be able to understand how these will affect the materials we use to make the reactor. That is complex science.'

Do the people at Culham believe that the private sector firms are hot on their heels? Chapman doesn't seem to believe so: 'We are still at the stage of trying to demonstrate it is possible to get more power out than you put in, and that it can be done on a commercial scale... 'Private industry is still some way behind that, as far as we can tell.'

JET was responsible for the 16MW fusion power record that helped to convince the world's governments that fusion was undoubtedly the energy source of the future. Amazingly, this was achieved in the 1990s when computers and the internet were still in their infancy and incapable of predicting or simulating the behaviour of hot plasmas.

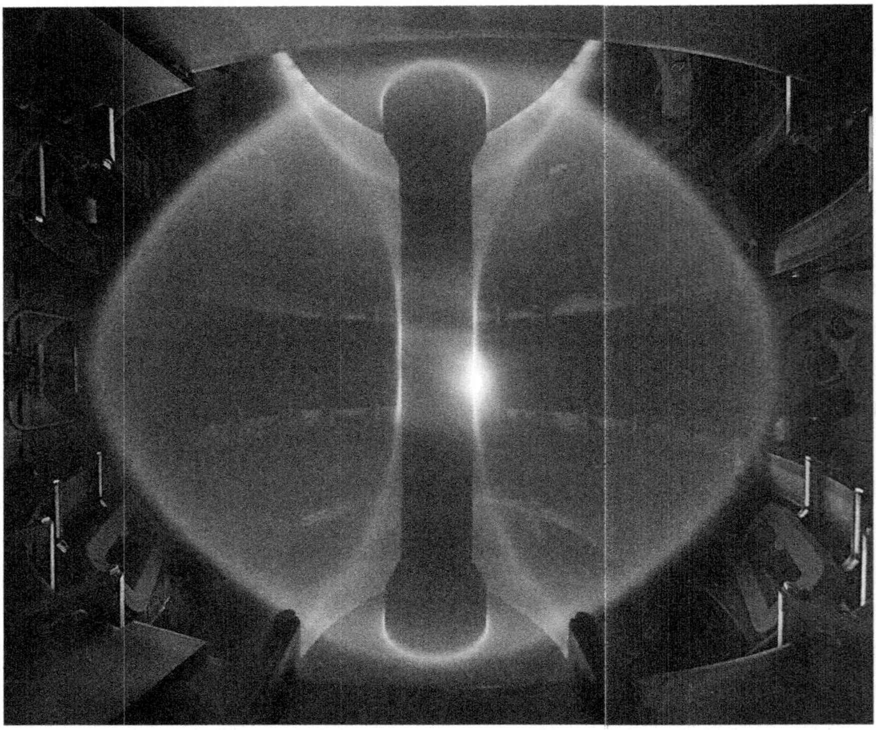

CULHAM CENTRE FOR FUSION ENERGY

Today the developments at Culham are marvellous and alien in their brilliance and sophistication. Unfortunately, they are still in the hands of governments. Following Brexit, JET has been thrust into a period of uncertainty as it is a joint European project. Scientists are worrying for their jobs and funding is again uncertain. It would be extremely unfeasible to move JET to Europe.

There is another small Eurotom reactor in the UK in the next-door building, down a few corridors which could be arguably more exciting than JET as it is of the very latest breed of spherical Tokamaks.

EUROFUSION

It is a £40 million upgrade of an earlier British ST. MAST (Mega Ampere Spherical Tokamak) is about to turn on for the first time.

The primary function of MAST is to prototype parts for ITER but it will also be available to UK universities. Prior to its upgrade, MAST has run over 30,000 plasmas and produced world leading data, providing cutting edge insight into the future of tokamak development.

The two reactors at Culham have created a talent pool and make the UK one of the best places in the world for fusion research. ITER is not really going to contribute so much to future reactor development as it will be demonstrating fusion works once and for all to the media and society.

Being a 1990s design, it is expected to work. It will undoubtedly be a fantastic experimental facility for fusion power stations but inescapably, a pressing task will be to learn how to make it smaller. In this way, findings made at MAST could be far more relevant to world fusion research than those made at ITER.

The remote handling facilities at Culham are quite astounding. Scientists are able to perform sensitive maintenance remotely on reactors using advanced robotics. Engineers can replace neutron damaged tokamak wall tiles, feeling any resistance through the controls so as not to damage the walls when screwing in new parts.

When witnessing these tasks being performed, it is hard not to be struck by the immensity of this technology and yet how little about it is known by the general public.

Since 2008 MAST and the MAST upgrade have been developed under the highly-respected Professor Steve Cowley who has been kind enough to afford me his time on a few occasions. Steve is a world class expert, a true child of fusion and theoretical physicist supreme. His successor, Professor Ian Chapman took over as the new CEO of the UK's Atomic Energy Authority in 2017, Professor Chapman has also been good enough to personally show me the latest updates at MAST. The UK AEA operate as much as is feasible, an open door policy to the public. For people looking to learn about nuclear fusion in the UK, the facilities and expertise available at Culham are unmatchable.

CHAPTER 7

OTHER NOTABLE TOKAMAKS

JT- 60 (Japan Torus 60)

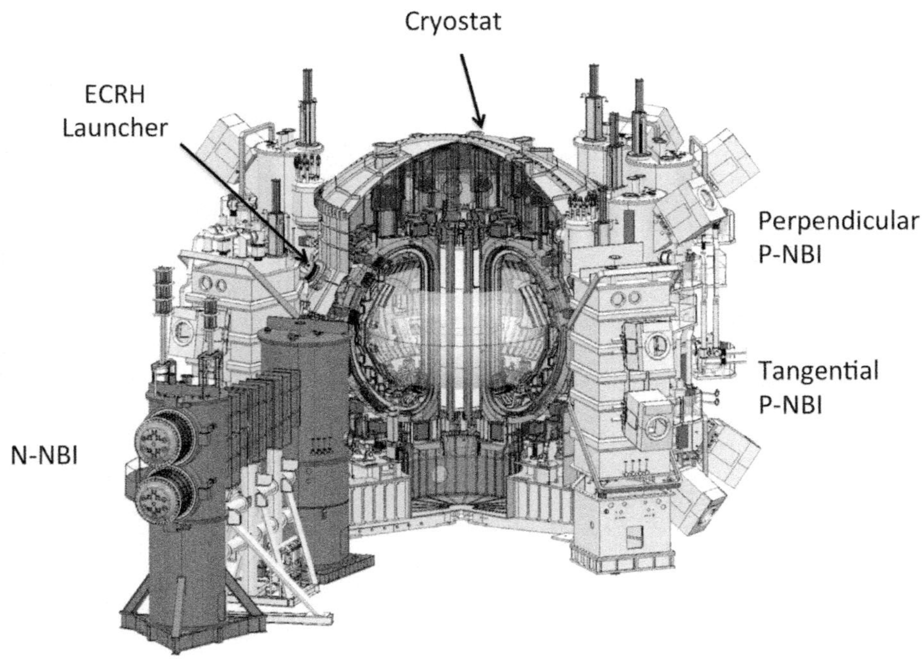

The JT-60 is a D shape tokamak similar to JET. It currently holds the record for the highest value fusion 'triple product'. This relates to the three most important properties a plasma must have in order to achieve an energy output of Q or higher.

These 3 quantities are:

n	**the particle density of a plasma**
t	**the confinement time**
T	**the temperature**

If these quantities all meet their criteria, net energy gain fusion will occur in the tokamak.

The JT-60 is also used to contribute towards ITER research. The Japanese fought strongly to have ITER built in Japan.

JT- 60 (Japan Torus 60)	
TYPE	Tokamak
OPERATION DATE	1985 - 2010
SIZE *(major radius / minor radius)*	3.4m *(11ft)* / 1.0m *(3ft 3in)*
PLASMA VOLUME	90m^3 *(3,200 cu ft)*
MAGNETIC FIELD	4 T *(torroidal)*
LOCATION	Ibaraki Prefecture, Japan

NSTX and NSTX - U at PPPL

The USA, PPPL- the Princeton Plasma Physics Laboratory has been home to some important fusion reactors. The TFTR (Tokamak Fusion Test Reactor) from 1982 - 1997 which produced 10.7 MW of power from a D-T plasma. And the NSTX (National Spherical Torus Experiment) which received a $94m upgrade (NSTX-U) that was completed in 2015. The NSTX reactors are notable for their ability to produce a high 'bootstrap' electric current.

This means that when the plasma pressure is raised, the tokamak is able to better handle the electrons that travel around the edge of the plasma, holding the plasma together improving stability and reducing heat loss. Electrons can become 'magnetically trapped' at the edge of the plasma in a tokamak and interfere with other electrons in motion. Information accrued at NSTX -U will be important for ITER and future torus reactors.

NSTX and NSTX - U at PPPL	
TYPE	Spherical tokamak
OPERATION DATE	1999 - Present
SIZE *(major radius / minor radius)*	0.85m / 0.68m
MAGNETIC FIELD	3 T
HEATING	11 MW
PLASMA CURRENT	1.4 MA
LOCATION	New Jersey, USA

The $94m National Spherical Torus Experiment Upgrade.

EAST - Experimental Advanced Superconducting Tokamak

The EAST reactor (Experimental Advanced Superconducting Tokamak) in China, has had a lot of press recently after achieving the record confinement time at temperatures more than twice as hot as the core of the Sun.

First plasma was in 2006 achieving an impressive 3 second plasma but by 2015, EAST had doubled that to over 6 seconds but in 2016, scientists recorded a 102 second confinement time at over 50 million degrees. Most importantly, this reactor was constructed for a mere £25 million which is a third of the cost of the NSTX- U and half the price of the MAST Upgrade.

Having run up against the problems discussed above when it comes to solar power, China has become a major player in nuclear fusion. Beyond ITER, of which it is a contributing member, the Chinese government is currently employing approximately 3,000 scientists and engineers in its national nuclear fusion programme. And it is making significant headway: in 2018, Chinese physicists achieved plasma temperatures of 100,000,000oC - a world first. And in 2017, the Experimental Advanced Superconducting Tokamak (EAST) became the first reactor in the world to sustain fusion conditions for 100 seconds.

China aims to be producing nuclear fusion energy by 2050: 'We are hoping to expand international cooperation through this device (EAST) and make Chinese contributions to mankind's future use of nuclear fusion,' Song Yuntao, a top official at the Anhei tokamak in Anhui province, recently told the press. The national government is already building a new fusion reactor, which is planned to supply commercial electricity.

The Experimental Advanced Superconducting Tokamak in Hefei, China

EAST - Experimental Advanced Superconducting Tokamak	
TYPE	Tokamak
OPERATION DATE	2006 - Present
SIZE *(major radius / minor radius)*	1.85m / 0.45m
MAGNETIC FIELD	3.5 T
HEATING	7.5 MW
PLASMA CURRENT	10 MA
LOCATION	Hefei, China

These developments are hugely exciting, but they have opened up international faultlines. 'We are very careful with intellectual property rights because in certain areas we are further along than the Chinese. We are very careful in what we share with China, and also Korea and India. We don't want to be in a situation where we have to buy the fusion of the future from the Chinese,' says EUROfusion's Donné.

Fusion for Energy (F4E), which is the EU's umbrella organisation to coordinate Europe's work on ITER, is also circumspect. Johannes Schwemmer, F4E's director, puts it simply: 'They [the Chinese government] are open to collaborating, but we should not forget that this is also a race to enable our science and industry in Europe to build this technology. We are not only doing this for all of mankind, but also for Europe. We want to be able to have the ability to build these machines, and not rely on China.'

A flattened view on the inside walls of Alcator C at MIT

Alcator C-Mod

The Alcator C- Mod (Alto Campo Toro, High Field Torus) is located at MIT's Plasma Science Fusion Centre in Massachusetts. It is a 3rd generation reactor, successor to Alcator A and B. Work began on Alcator C in 1976. In 2016, it achieved the highest plasma pressure and magnetic field strength *without the use of superconducting magnets.*

Superconducting magnets need to be cooled to temperatures well below zero degrees and closer to zero kelvin (-275 degrees) to maximise their performance. This requires the use of liquid nitrogen or helium. Superconductors are expensive and add significant complexity to the tokamak making them an inelegant feature. Alcator C has achieved great results without them.

ALCATOR C-MOD	
TYPE	Tokamak
OPERATION DATE	1991 - 2016
SIZE *(major radius / minor radius)*	0.68m / 0.22m
PLASMA VOLUME	1m^3
MAGNETIC FIELD	3 - 8 T *(torroidal)*
PLASMA CURRENT	0.4 - 2.0 MA
LOCATION	Massachusetts, USA

DEMO (The DEMOnstration Power Station)

ITER, is only an experimental reactor. It was not designed to power the gird. Before an actual power station can be constructed, ITER must demonstrate its ability to confine and control high density plasma without destroying the walls of the reactor. Upon satisfactory results, governments will reconvene to discuss the DEMO concept. DEMO, (The DEMOnstration Power Station) was planned in detail in 2009 by the Atomic Energy Authority in the UK and Euratom. It is intended to be the first fusion reactor that will use the heat generated from the neutron emission from the plasma to drive a steam turbine and generate electricity.

Let's assume for a moment that of the hare and the tortoise in this race, it's the slow and steady ITER that triumphs. Much of the world's scientific community believe that this government-backed mega-project is the only option with the resources to prepare the way for fusion electricity. But ITER will not supply a grid; rather, its successor, DEMO, which is already in the planning, will become the world's nuclear fusion power station. That's the plan anyway.

ITER's giant tokamak reactor is scheduled to start running in 2025, with experiments then proceeding over the course of the next 20 years. Scientists and engineers are already envisaging construction on DEMO starting during this period. By 2050, DEMO should be producing electricity.

Both DEMO and ITER will use tokamak reactors, but on completely different scales. While ITER is currently expected to be the largest tokamak ever built, DEMO will be huger still, by some 15 per cent. So DEMO's vacuum chamber will have a volume of $2200m^3$ - dwarfing ITER's $800m^3$.

This is necessary in order to scale up the amount of energy that ITER scientists believe is required to make fusion electricity production possible. According to Tony Donné, programme manager at EUROfusion, one of the many agencies overseeing ITER, the extra capacity of DEMO is crucial: 'We call the ratio of 'power out' to 'power in' the Q factor, and at the moment no fusion reactor has exceeded a Q factor of one (Q1). ITER's performance we call Q10, but because of the inefficiencies inherent in generating electricity – with heat exchange and turbines – Q10 is nowhere near enough to be useful. We need DEMO to produce about Q25, somewhere around 2 to 4GW of thermal output. Possibly full-scale commercial fusion stations will need to produce an even higher Q factor, maybe Q100.'

Nowhere could the contrast be greater between public and private sector approaches to fusion electricity than in the scale of DEMO. Private companies, aiming to use small machines using nimble technology, are seeking to manoeuvre their way to fuelling the grid in about ten years from now. Meanwhile, ITER scientists don't believe it is possible until ITER has run its course and prepared the ground for DEMO in 2050. Who knows which side will be proved right; ultimately, does it even matter? Surely the aim is nuclear fusion itself, however it is achieved. Our planet, with a growing population that is sapping away our resources, needs a clean alternative to fossil fuels, and soon. Otherwise, all that awaits us is the catastrophe of climate change.

It is very important that the plasma facing walls of the reactor are extremely durable when it comes to power stations. If the reactor needs to be constantly dismantled for wall replacement, it will delay the power production process making the reactor less efficient and less desirable. As the neutrons bombard the reactor walls, not only do they damage them, but over time, the walls become radioactive. This is not the same long-lived radioactivity as fission waste, which must be buried for thousands of years, but it still means people will need to wear protective clothing and robots will be used to change out the damaged wall plates. Again, this adds significant expense and maintenance delay. Better plasma control will increase the life of the walls, as will finding advanced materials that can cope with the neutron bombardment.

In Germany, concepts for liquid metal walls are being researched. This could save having to change out the walls, but is perhaps another overly complicated process for power stations to have to cope with.

Advanced materials created by modern 3D printers may be able to create wall plates made from a mixture of metals chosen for their durability qualities. These 3D printers often create disordered structures which may be better placed to handle neutron bombardment. In short, the longer the walls can last, the more efficient the power station will be.

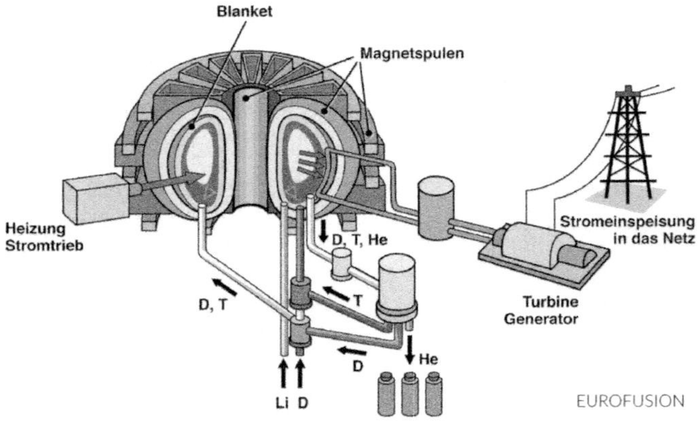

EUROFUSION

The used wall parts that have become radioactive will have to be disposed of in a similar way to fission but for one hundred years as opposed to thousands. It is a process that we have mastered thanks to fission waste, so this will not require new innovation. However, there could be methods of further reducing the short-lived radioactive waste of tokamak wall lining.

Another factor to be considered with using Deuterium -Tritium fuelled Tokamaks as a power station is the supply of tritium. Deuterium is not in short supply, we can find it in relative abundance in seawater but tritium occurs naturally on earth in very small amounts. Currently the world demand for tritium is 400 grams per year. Scientists 'breed' it in laboratories. This can be done by bombarding Lithium 6 with a neutron, the product is helium and tritium. Tritium currently has uses in medical equipment; and gun sights that glow in the dark to name a few; but it is expensive. $30,000 per gram as at the time of writing. A power station would require 300g of tritium per day to generate 800MW of power so a new way of procuring tritium is required. Sometimes as heartbreakingly complex as fusion can seem to be, it also has its brilliant conveniences. The solution for a tritium source for Tokamaks is for them to *breed their own* tritium.

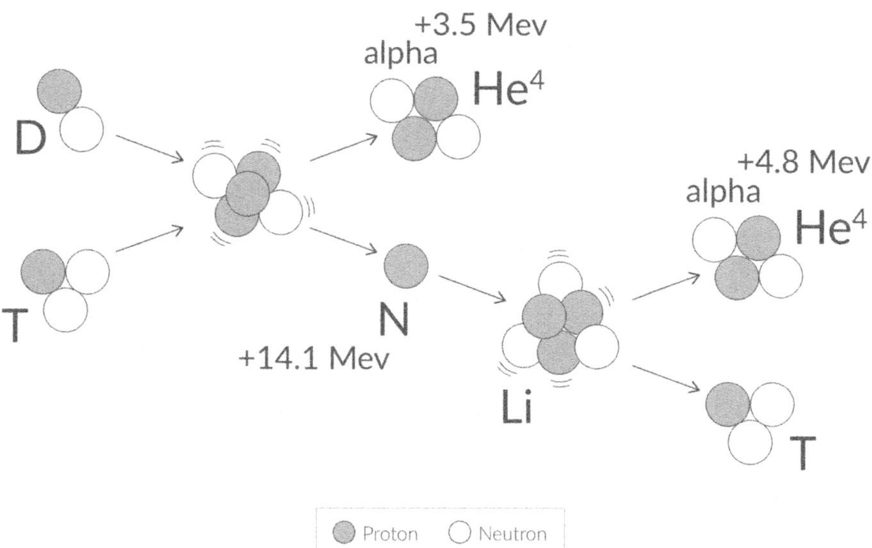

Tritium created as a product of lithium 6 and a neutron in a tokamak.

A tokamak is after all, a neutron source. DEMO is designed to have a 'lithium blanket' around the outside of the vessel so when the neutrons from the plasma pass through the walls, they interact with the lithium and create more tritium fuel.

Technologies to be incorporated into the DEMO tokamak will not be finalised until data starts coming in from ITER. Super conductors and liquid walls, divertor positioning and improved vessel shapes will all be considered when designing DEMO. The conceptual design was due to be completed by 2017 and the engineering design by 2024 but as ITER has been delayed again, so has DEMO. The current roadmap as set out by the UK AEA and Euratom in 2004, suggests a fully functioning electricity generation demonstration will not take place until 2048.

That is for a technology that was successfully demonstrated in 1997 by scientists working prior to the arrival of supercomputing or even conventional PCs. A technology of such significance as able to bring humanity into a new age:

The Fusion Age.

Private Tokamaks

The very fact that private firms and venture capitalists have started to plough money into nuclear fusion shows that the dream that has been out of reach for so long is now a potential reality. In the venture capital world, there is an unwritten rule that you don't touch projects that will take more than a decade to deliver; so the injection of these very funds into this area shows a growing confidence that nuclear fusion is deliverable sooner than many may think. Today's investor is putting money where his or her mouth is.

PULSAR FUSION

What is interesting about the many businesses now dedicated to the pursuit of nuclear fusion energy, is that the physics essentially has not changed. Since ITER's inception, it is engineering developments that have led the way, and the private sector is characterised by an 'engineering first, physics second' approach. Since the Second World War, it is private companies who have led the way in commercialising technology and engineering innovations: Microsoft, Apple, Tesla. Investors believe the same is possible with nuclear fusion. Whether the aim of achieving fusion in itself is realised, or if it is simply a matter of investigating the developments along the way, there is money to be made.

You may be able to tell by now from reading this book, I am not very happy about waiting until 2050 to see nuclear fusion take over, and I'm not alone.

I have learned over my career that large enterprises are less efficient than small ones and governments are less so by a factor of 100. In 1997, governments were the only entity with access to the resources that were needed for fusion development. Now, small enterprises can share workspace and expensive machinery, prototyping facilities are cheap and multi-million-dollar super computer clusters can be harnessed for a small monthly fee. Scientists are no longer obliged to contract with huge government projects like CERN or ITER and are now less incentivised to do so.

A private company may be able to offer them more freedom of research, more money and the ability to put their research into action in their lifetimes. Companies do not have to follow government mandate, contrary to popular public belief. They are not vulnerable to political events like Brexit and they can change direction on a hairpin if a project no longer makes sense. (And they will be required to, if they are to survive).

It is my belief that very recently a tipping point has been crossed. Private enterprises are not only just as equipped as national projects but now better placed to realise their ambitions.

If there was ever an ambition to test this theory it is with nuclear fusion. The most complex and revered technology of them all. Many people have rolled their eyes at me for this but it is already happening. I visited the ITER conference in Marseille in 2015 and sat and observed the updates from all the contractors who are making the many, many components of ITER. Unsurprisingly, all the components that will build the giant reactor are built by companies, many of which are small, specialist firms who are, at the expense of ITER, becoming World leaders in creating tokamak components.

They are not under any non-compete agreements and keen to make money just like any typical business. The real problem is they are all supplying one of the largest multi-national projects in the World and, as always in such cases, there are far, far too many people involved in it. People make problems and when the governments come up against problems they hire more people. Anyone who has ever run an organisation of any kind would put their head in their hands at a set-up such as this. So, all the parts are created by conventional specialist companies, it's just the assembly and organisation that is being handled by 7 respective nations.

The really frustrating part is that none of this relates to the actual science of fusion. Instead it is the science of multi-government mega project management that is stultifying progress.

There is no component of ITER, or any other fusion reactor for that matter, which a private company could not equally get hold of or manufacture. Accredited and experienced scientists are more than willing to jump ship if the equipment is there, the vision is sound and the funding is in place.

The race to fusion is an exciting one, as now supercomputing is predicting reactors can be a fraction of the size, coupled with growing demand for smaller 100MW reactors. It ultimately all comes back down to our one equation of fusion.

Funding = Fusion

It turns out the most important skillset for fusion science is not actually plasma physics after all. It is the ability to raise the finance.

Leading the UK effort: Pulsar Fusion, Tokamak Energy and First Light Fusion

Just across the Channel from the ITER behemoth are a number of young, dynamic nuclear fusion enterprises.

Tokamak Energy and First Light Fusion are key figures in the thriving Oxfordshire scientific community. A spin-off from Oxford University, First Light Fusion is pursuing a radical path to nuclear fusion, in which a projectile is fired to create 'captivation bubbles' in the tritium-deuterium fuel. As the bubbles collapse, calculations suggest that the deuterium and tritium within them will be compressed to such a degree that fusion occurs. To date, this has not achieved all three conditions needed in order to achieve nuclear fusion, but the highly academic and agile team believe they have 'a clear vision of the pathway to fusion'. Tests begin in 2019 to see whether their method can yield fusion.

Tokamak Energy is one of the oldest UK firms in the business, having been formed in 2009 out of the Culham Centre. Their mission, to find 'a faster way to fusion', is well underway: Tokamak Energy has already built three spherical fusion reactors which use high temperature superconductors to heat the plasma and contain it in an extremely strong magnetic field. The small reactors are more flexible and cheaper to build, giving the company more variables to test. In June 2018, Tokamak's ST40 reactor achieved plasma temperatures hotter than the centre of the Sun - over 15,000,000oC. Jonathan Carling, their chief executive, believes that the company will be producing net energy gains by 2022 and 'supplying energy to the grid by 2030.'

Pulsar Fusion, who possess the largest private facility at the time of writing, formerly known as Applied Fusion Systems, is the brainchild of the author.

Dr James Lambert - head of operations, says that using 3D printers to make their components and supercomputers to simulate plasma behaviour gives Pulsar Fusion a key technological edge: 'Plasma behaves in unpredictable ways,' he told the BBC in 2017. 'Now that the price of supercomputing has plummeted, it is giving fusion researchers the ability to test their ideas before building them.'

Pulsar aims to build several small-scale reactors at their facility in Bletchley, based on the experimental Mega Amp Spherical Tokamak (Mast) at the Culham Centre for Fusion Energy but designed to yield breakthroughs in propulsion and fusion innovations in Space rather than grid power. However, with the use of advanced rapid prototyping techniques, Pulsar scientists aim to slash costs and work quickly and dynamically to solve problems, in a way the vast public projects like ITER simply can't. For Pulsar Fusion, it's not just about achieving energy-gain fusion, but about positioning themselves for a post-fusion world: 'When ITER works, the appetite for fusion will be huge," says Lambert. 'It will be the biggest innovation of our generation and it is going to be companies that are starting in the fusion business now that others will come to.'

So, whether the aim is to position a company as the go-to option once nuclear fusion energy is achieved, or to be the first to deliver it to the grid, or to try cutting-edge paths to nuclear fusion that are less tested, the UK has a wide field of private firms looking to make nuclear fusion a reality, and gain from it. They are increasingly hot on the heels of ITER.

PULSAR FUSION

The US:
Where big dreams meet even bigger philanthropy

It is no secret that in a country of small government like the US, it is the private sector that has so often led the way in innovation. Whereas in Europe, governments have often taken the lead in funding the technologies of the future, in America, philanthropy from the super-rich fills the gap. Disruptive companies such as Amazon, Apple, Paypal and Microsoft have changed the face of society in untold ways. And now, their founders are ploughing their philanthropic efforts into nuclear fusion. Firms like General Fusion, TAE Technologies, Helion Energy and Commonwealth Fusion Systems have been funded by the likes of Amazon's Jeff Bezos, Microsoft founders Bill Gates and the late Paul Allen, and Virgin's Richard Branson. With both the Obama and Trump administrations ambivalent about nuclear fusion, startups have sprung up across the country.

Like in the UK, all these ventures are focusing on smaller experiments using new technology. The billionaires backing these enterprises are all clean-energy advocates. Christofer Mowry, CEO and director of General Fusion, recently told Bloomberg News: 'If you care about climate change you have to care about the timescale and not just the ultimate solution. Governments aren't working with the urgency needed.' These firms are aiming to achieve nuclear fusion energy supplied to the grid way ahead of ITER and DEMO's forecasted 2050.

Mowry's **General Fusion**, which is backed by Bezos, is a Canadian-based firm found in Burnaby, just outside Vancouver. Using extreme compression, or 'magnetised target fusion', the firm claims it has discovered 'the most practical path to fusion energy'. Their 3 metre wide fusion reactor looks like a sea mine: spherical and covered with spikes which hold some 220 pistons. Controlled and synchronised by computers, these pistons fire simultaneously into the chamber, using anvils to send a powerful shock wave through the plasma. By using compression, the fuel's density is pushed up a thousandfold, which increases its temperature from 5,000,000°C to 150,000,000°C.

This creates the conditions for fusion to take place. Mowry plans to have a demo plant up and running by 2023, and after that, to be bringing power to the national grid.

TAE Technologies, which was bankrolled for many years by the late Paul Allen, who co-founded Microsoft, is a California-based enterprise with a radical approach to fuel. Once known as Tri Alpha Energy, the company was a lone wolf in the nuclear fusion industry when it was founded over 20 years ago. Now, it has a couple of dozen competitors. TAE Technologies, who pride themselves on producing a brand of 'friendly fusion', have named their 25-metre long cylindrical reactor Norman, after their late co-founder, Dr. Norman Rostoker. Rostoker, who was a University of California plasma physicist, was at the intellectual heart of TAE until his death in 2014.

Rather than using hydrogen isotopes in Norman's reaction chamber, TAE Technologies are working with a form of boron instead. In the reactor, which has been operational since 2017, beams of high-energy plasma are fired at each other simultaneously, travelling at speeds of approximately 1,000,000 kilometres an hour. When they meet, they form a spinning, cylindrical cloud which is then kept stable by beams of deuterium, which are fired into the cloud from outside the chamber.

Norman has, to date, maintained plasma at temperatures which are two low and for timescales that are too short to be a workable option for a net energy gain reaction. However, TAE hopes to drive these figures up and to make a hydrogen-boron fuel a viable alternative to deuterium-tritium fusion. Why? In order to make fusion even safer. As we know, the reaction produced by deuterium and tritium together in a tokamak produces phenomenal amounts of energy. As the energy is released, about 80 per cent of it becomes kinetic energy in the form of neutrons moving away from the reaction. Proposed designs for fusion power plants at present are looking at finding a suitable material to absorb these neutrons and turn their kinetic energy into heat, which could then produce steam to turn a turbine. In this way, electricity could be generated. Scientists are considering the possibility of using lithium for this purpose, which would have the added bonus of producing new tritium, thereby generating further fuel for the power station.

The reason that TAE Technologies are looking for an alternative to this is that a tokamak reactor working using these methods would become radioactive. As the whole reactor absorbs the neutrons, the structure will degrade and become damaged, producing significant maintenance costs. Plus, there are the important environmental implications to consider. While we're not talking about radioactivity levels anywhere near as high as those found in a nuclear fission reactor, nuclear fusion is supposed to be clean and safe. By pursuing hydrogen-boron as fuel, TAE Technologies are keeping things as clean as possible. It is also a highly efficient model: while deuterium-tritium fusion loses energy at each step of the process, hydrogen-boron reactions can directly create currents in external conductors, without needing to use heat.

However, this path is not an easy one to have selected. TAE Technologies' hydrogen-boron fusion will take temperatures in the order of billions degrees Celsius for fusion to happen. This kind of temperature has never been achieved in any fusion experiment. Nonetheless, TAE scientists are confident that they will be supplying the grid with electricity by 2030.

Commonwealth Fusion Systems (CFS) was founded in 2017 by six MIT academics and is easily one of the most ambitious projects in the field. Its aim? To be 'the fastest path to clean, limitless fusion energy.' CFS is characterised by an urgency in its desire to achieve nuclear fusion energy: the Cambridge, MA venture wants fusion energy delivered to the grid, yesterday. As their website argues: 'Energy use is a fundamental enabler for improved quality of life worldwide, yet historically it has come at an extreme cost to the environment. Fusion energy can change that. We aim to deploy fusion energy to the world on a timescale that will make a difference.'

CFS is backed by a number of the super-rich: Bezos, Gates and Branson have all donated. Oil producers who are future-proofing themselves against a carbon-neutral planet have also chipped in, with Eni SpA, the Italian energy giant, providing $50 million in March 2018.

It has also secured funding from Breakthrough Energy Ventures, a fund that draws together a number of conservation-minded philanthropists. Featuring some of the truly eye-wateringly rich, their list of investors is like a business hall of fame: Mark Zuckerberg, Bill Gates, Jeff Bezos, Richard Branson, Michael Bloomberg and George Soros, to name but a few. Breakthrough Energy's goal is 'to make sure that everyone on the planet can enjoy a good standard of living, including basic electricity, healthy food, comfortable buildings, and convenient transportation, without contributing to climate change. Our strategy links cutting-edge, government-funded research to patient, risk-tolerant capital so that more clean energy innovations get to market faster.'

The aims of CFS and Breakthrough Energy coalesce nicely around each other, and it is easy to see why the MIT professors might attract this funding: both organisations want clean energy, fast. As Commonwealth Fusion's CEO Bob Mumgaard told Bloomberg News last year: 'The greater danger is not having anybody succeed than having everybody... We need more smart people driving very hard to crack this.' CFS plan to have a prototype for an energy-gain reactor in production by 2025. No-one is driving harder.

Helion Energy is another major player in the US and is attracting growing numbers of funders, both public and private, including PayPal's Peter Thiel. Having grown frustrated with the likes of ITER's progress - or perceived lack of it - founding partners Dr. David Kirtley, Dr. George Votroubek and Chris Pihl established Helion. With years of experience in fusion between them, the team are seeking to use the latest physics and cutting edge engineering to make a fusion generator '1,000 times smaller, over 500 times cheaper, and realizable 10 time faster than other approaches.'

Kirtley's team have focused on the end goal - the production of electricity - rather than on the means of getting there. By taking this unique approach, the Helion team have used magneto-inertial fusion to produce a commercially viable solution, thanks to lower costs than some of its competitors. By employing magnetic compression techniques, they have eliminated the need for pistons, lasers or beams, all of which are expensive and use energy in and of themselves. Helion's model converts energy directly, making it cheaper and more efficient. To date, Helion scientists have achieved fusion temperature plasmas with compression fields in excess of 100 Tesla. They hope to bring fusion electricity to market at a cost of 4 cents per KWH - less than a third of the current average price in the US. Their efficient, cost-saving approach is yielding results and winning them a string of awards.

Having watched government-run programmes, such as ITER, stall and blow the budget, these private ventures are seeking an alternative. As physicist Freeman Dyson argues: 'The existing fusion program is leading to a huge source of centralized power, at a price that nobody except a government can afford' (Dyson, 1998: 40). The private sector is doing everything it can to provide an alternative narrative. They want fusion sooner, they want it to be deliverable, and they want it to be affordable. Time will tell as to whether they can out-strip ITER and finally bring fusion energy to the market.

The giant thermonuclear reactor in the sky

It's worth noting that not everyone is backing nuclear fusion. Elon Musk, co-founder of PayPal, SpaceX, Tesla and SolarCity, says we should be focusing our energy on that produced by the Sun: 'We've got a giant thermonuclear reactor in the sky... It shows up every day very reliably. If you can generate solar panels and store it with batteries, you can have it 24-hours a day.' For Musk, the priority is in harnessing the Sun's energy, rather than trying to replicate it on Earth.

In countries such as India and China, huge solar parks have been springing up as these hugely populous countries try to keep up with energy demand in a clean way (approximately two thirds of China's electricity is still generated by burning coal). With the world population projected to rise to 9.8 billion by 2050, these countries have been ploughing significant money into solar power. Indeed, China produces more solar energy than any other country in the world, with an enormous national capacity of 130 gigawatts. 1,500 megawatts is generated at one solar plant alone: the largest of its kind in the world currently sits in China's Tengger Desert.

But solar power runs into problems. First, huge solar plants need an enormous amount of land to be built on. Given that many of the largest users of electricity are countries with large, sprawling populations, this isn't easy. Add to that the fact that solar power loses capacity as it travels over transmission lines, and the problem gets even trickier: solar farms need to be built close to where their resultant energy is used. In most developed countries, this just isn't possible, due to lack of space and planning issues. So, China's solar plants, which are generally housed hundreds of miles from the cities that use them, produce masses of wasted energy: just one sixth makes it into Chinese households, on average. The rest just leaks out of transmission lines.

And while solar power is billed as the clean option, we will soon have to turn our attention to the lifespan of solar technology. On average lasting just 30 years, these solar plants are increasingly going to be churning out defunct panels, which contain harmful chemicals such as sulphuric acid. Hard to recycle, from 2040 onwards China and India will be presented with the enormous task of disposing of millions of these panels. While they may not contain chemicals that are toxic in the way waste from nuclear fission is, at present the world awaits a plan for what to do with the detritus of this 'green' technology.

And so it is that so many of us return to nuclear fusion; at present, it represents the cleanest and most efficient way of producing energy that can meet the world's growing energy needs. General Atomics estimate that you would need 755 acres of solar panels to generate the electricity yielded from just 11 pounds of hydrogen. It's a no-brainer.

CHAPTER 8
THE FUTURE

Helium 3 and Aneutronic Fusion Research

Once ITER does successfully demonstrate a 01- 15 times net energy gain the world media will fall upon Nuclear Fusion. The major questions will undoubtedly be: How can we make it cheaper?

In short, the first obvious route is to reduce the immense size of it. Reactors need to be significantly smaller than ITER to become cheaper. This will require careful diagnostics and analysis of the data produced at ITER before governments commit to DEMO. We will need the help of ever progressing modern computer simulation capabilities and vitally, the further understanding of plasma behaviour. One thing is for sure, in 2025, there will be a lot more money in fusion.

The second factor to making the reactor cheaper is to increase the life of the vessel. The Deuterium - Tritium fusion reaction releases its energy in the form of neutrons that bombard the reactor walls. Their kinetic energy produces the heat that is then in turn used to drive a steam turbine.

Aside from steam turbines being enormously costly apparatus, the neutrons damage the reactor walls as they smash into them. Conventional metals have lattice like structures that depend on their uniformity for structural stability. The neutrons fly straight into the micro structures of the metals and ultimately the walls become radioactive and structurally unstable over time. This means expensive robotics must be employed to change out the lining of the vessel walls. This is time consuming and will keep the reactor offline for long periods of time preventing it from doing its job of generating energy. Luckily the energy created in fusion reactions is so enormous these setbacks are not significant enough to prevent fusion taking over as the world's dominant energy source.

The most obvious solution to all of this is to swap tritium for helium-3. When helium- 3 is fused with deuterium, rather than a neutron being the expelled particle, it is a proton. A proton is a charged particle so not only can it be controlled by the tokamak's magnetic fields, but electricity can be captured as direct charge from the protons. This means one could harness electricity directly off the reactor without the need of a steam turbine and the reactor walls would be spared the neutron bombardment increasing their service time by multiple factors. There would still be occasional neutron created as inevitable deuterium - deuterium reactions occur, but as DD fusion requires higher temperatures, there would be far fewer to a point of insignificance.

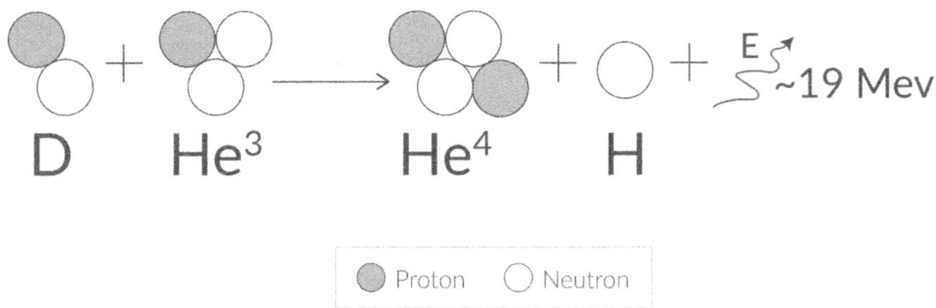

This would leave us with a longer lasting reactor without the need for a steam turbine. There are setbacks to overcome for this to be achieved though. Helium-3 - deuterium fusion requires slightly higher temperatures than deuterium - tritium but produces a little more energy. The major obstacle though is helium-3 doesn't occur naturally on earth. Nor does tritium but we can quite easily breed that using lithium. DEMO is designed to have its own tritium breeding belts so it can create its own fuel supply. Helium-3 however is harder to come by on earth.

Our moon has an abundance of helium-3 in its regolith which is the term for lunar soil. So rich in fact that missions to the moon have been considered by many. One ton of helium-3 would power the United States for a year giving helium-3 an economic valuation of $3 billion per ton which makes a moon mission almost viable. The helium-3 however, is engrained in the lunar soil and would need to be harvested a volume of hundreds of tonnes. That would require the use of a combine harvester type machine which at present would be a challenge to get to the moon. But then perhaps by that time fusion power will have given us nuclear thrusters capable of significantly reducing the costs of launching heavy equipment into space.

Additionally, people are protective of the moon and its romantic associations. Harvesting would need to be done on the dark side of the moon to prevent public opposition and then there is the debate of who owns the helium-3.

The fusion story certainly does not end with DT fusion and the development of anatomic fusion will no doubt be prompted following the success of ITER.

Fusion Thrusters and Super Innovations

Fusion energy may well be the most important technological innovation of my lifetime and if you are reading this book, perhaps yours too. I personally see it making more of a significant change to human lives than the invention of gunpowder. But for good. Fusion technology development will bring about a plethora of propulsion, defence and weaponry innovations. NASA have even started looking at using a spherical torus fusion reactor as a space propulsion vehicle concept for fast interplanetary travel. Which in turn, could even bring about the side effect of relativistic time travel.

Winston Churchill quoted in his book – ***My Younger Years***
"Everything I was sure or taught to be sure was impossible, has happened."

ABOUT THE AUTHOR

Richard Dinan is the founder of Pulsar Fusion, a private manufacturer of nuclear fusion reactors in the UK and author of 'The Fusion Age - Modern Nuclear Reactors' book.

Richard retains an interest in a number of other technology businesses. His companies have built and designed rapid prototyping facilities for European Defense and Aerospace companies including Airbus, Jaguar-Range Rover and The Dubai Space Agency. His robotics and prototyping innovations have been endorsed and showcased by the British Government and have been used to represent British innovation by David Cameron in Germany in 2014.

Richard has been a guest lecturer at Imperial College to their (MSc) - Master of Science students on Business development and rapid prototyping. Richard has been a guest lecturer at Oxford University - On his filmed lecture - 'Private Nuclear Fusion Reactors' available on YouTube.

He currently serves on the judging panel of the Great British Entrepreneur Awards.

Printed in Great Britain
by Amazon

44978643R00061